OXFORD GEOGRAPHICAL AND ENVIRONMENTAL STUDIES

General Editors: Gordon Clark, Andrew Goudie, and Ceri Peach

SUSTAINABLE LIVELIHOODS
IN KALAHARI ENVIRONMENTS

ALSO PUBLISHED BY
OXFORD UNIVERSITY PRESS
IN THE OXFORD GEOGRAPHICAL AND
ENVIRONMENTAL STUDIES SERIES

Sustainable Livelihoods in Kalahari Environments

A Contribution to Global Debates

Edited by

Deborah Sporton

and

David S. G. Thomas

OXFORD

UNIVERSITY PRESS

OXFORD

UNIVERSITY PRESS

Great Clarendon Street, Oxford OX2 6DP

Oxford University Press is a department of the University of Oxford.
It furthers the University's objective of excellence in research, scholarship,
and education by publishing worldwide in

Oxford New York

Auckland Bangkok Buenos Aires Cape Town Chennai
Dar es Salaam Delhi Hong Kong Istanbul Karachi Kolkata
Kuala Lumpur Madrid Melbourne Mexico City Mumbai Nairobi
São Paulo Shanghai Singapore Taipei Tokyo Toronto

and an associated company in Berlin

Oxford is a registered trade mark of Oxford University Press
in the UK and in certain other countries

Published in the United States
by Oxford University Press Inc., New York

British Library Cataloguing in Publication Data

Data available

Library of Congress Cataloging in Publication Data

Sustainable livelihoods in Kalahari environments : contributions to global
debates /
edited by Deborah Sporton and David S. G. Thomas.
p. cm.—(Oxford geographical and environmental studies)
1. Range management—Kalahari Desert. 2. Range ecology—Kalahari
Desert. 3. Pastoral systems—Kalahari Desert. 4. Land use, Rural—
Kalahari Desert. 5. Rural development—Kalahari Desert. I. Sporton,
Deborah. II. Thomas, David S. G. III. Series.
SF85.4.K35 S88 2002 333.74′096883—dc21 2002020086

ISBN 0-19-823419-8

1 3 5 7 9 10 8 6 4 2

Typeset in 10 on 12 pt Times New Roman
by SNP Best-set Typesetter Ltd., Hong Kong
Printed in Great Britain
on acid-free paper by
T.J. International Ltd, Padstow, Cornwall

EDITORS' PREFACE

Geography and environmental studies are two closely related and burgeoning fields of academic enquiry. Both have grown rapidly over the past few decades. At once catholic in its approach and yet strongly committed to a comprehensive understanding of the world, geography has focused upon the interaction between global and local phenomena. Environmental studies, on the other hand, have shared with the discipline of geography an engagement with different disciplines, addressing wide-ranging and significant environmental issues in the scientific community and the policy community. From the analysis of climate change and physical environmental processes to the cultural dislocations of postmodernism in human geography, these two fields of enquiry have been at the forefront of attempts to comprehend transformations taking place in the world, manifesting themselves as a variety of separate but interrelated spatial scales.

The Oxford Geographical and Environmental Studies series aims to reflect this diversity and engagement. Our goal is to publish the best original research in the two related fields and, in doing so, to demonstrate the significance of geographical and environmental perspectives for understanding the contemporary world. As a consequence, our scope is deliberately international and ranges widely in terms of topics, approaches, and methodologies. Authors are welcome from all corners of the globe. We hope the series will help to redefine the frontiers of knowledge and build bridges within the fields of geography and environmental studies. We hope also that it will cement links with issues and approaches that have originated outside the strict confines of these disciplines. In doing so, our publications will contribute to the frontiers of research and knowledge while representing the fruits of particular and diverse scholarly traditions.

Gordon L. Clark
Andrew Goudie
Ceri Peach

ACKNOWLEDGEMENTS

Many individuals have contributed in various ways to the completion of this volume. First and foremost we thank all the contributors for their patience in seeing this book through to publication and for their thoroughness in preparing their final manuscripts. We would also like to acknowledge the considerable contribution of Paul Coles in the Department of Geography at Sheffield University, who has performed wonders converting our scribbles into the maps and diagrams that appear in the book. The idea for the book originated from a workshop held at the University of Sheffield on 'Sustainable Livelihoods in Marginal African Environments' that we convened in 1997 and several chapters are revised and updated versions of papers presented at this meeting. We would like to thank all those who participated at the workshop and to acknowledge the support of the Economic and Social Research Council Global Environmental Change Programme who funded both the workshop and some important components of the research that has led to some of the outputs within this book. This research would not have been possible without the efforts of Jean Morrison and Jem Perkins, whose collaboration and companionship made for some memorable field trips in the Kalahari. We must also thank those who eased various logistical elements of the research, notably Marty McFarlane and Paul Shaw and his family.

More recently our research in the Kalahari has been funded by the DFID Natural Resources Management Programme whose support is gratefully acknowledged. Chapter 3 in particular draws on some of the findings from the DFID-funded PANRUSA research project (http://www.shef.ac.uk/panrusa) in Botswana, Namibia, and South Africa. Our thanks go to Chasca Twyman for her unstinting support as research associate on the project, to Andy Dougill, Andre Van Rooyen, James Drummond, and to our other collaborators in southern Africa. We are also grateful to Eric and Karin Walker for their logistical and fieldwork support.

The authors and publishers thank the following for their permission to reproduce copyright material within this book:

Figs. 2.1 and 6.1 have been reprinted from D. S. G. Thomas and D. Sporton, 'Understanding the dynamics of social and environmental variability: The impacts of structural land use change on the environment and peoples of the Kalahari, Botswana', *Applied Geography*, 17/1: 11–17 (Copyright 1997), with permission from Elsevier Science.

Figs. 2.2 and 4.1 have been reprinted from D. S. G. Thomas and P. Shaw *The Kalahari Environment*, (Copyright 1991), with permission from Cambridge University Press.

Fig. 5.1 has been reproduced from A. J. Dougill and N. M. Trodd, 'Monitoring and modelling open savannas using multisource information:

analyses of Kalahari studies', *Global Ecology and Biogeography*, 8: 211–21 (Copyright 1999), with permission from Blackwell Publishers.

Figs. 5.4, 5.5, and 5.6 have been reproduced from A. J. Dougill, D. S. G. Thomas, and A. L. Heathwaite, 'Environmental change in the Kalahari: Integrated land degradation studies for non equilibrium dryland environments', *Annals of the Association of American Geographers*, 89: 420–42 (Copyright 1999), with permission from Blackwell Publishers.

Fig. 6.3 has been reproduced from D. S. G. Thomas, D. Sporton and J. Perkins, 'The environmental impact of livestock ranches in the Kalahari, Botswana: Natural resource use, ecological change and human response in a dynamic dryland system', *Land Degradation and Development*, 11: 327–41 (Copyright 2000), with permission from John Wiley & Son.

Figs. 7.1 and 7.2 have been reproduced from C. Twyman, 'Participatory conservation? Community-based natural resource management in Botswana', *The Geographical Journal*, 166/4: 323–35 (Copyright 2000), with permission from Blackwell Publishers.

Every effort has been made to trace and contact copyright holders. We apologise for any errors or omissions that may have been made and would be pleased to make any corrections at the first opportunity should this be necessary.

CONTENTS

LIST OF FIGURES

LIST OF BOXES

LIST OF TABLES

LIST OF CONTRIBUTORS

Andrew J. Dougill (adougill@env.leeds.ac.uk) is a lecturer in Environment and Development based in the School of the Environment, University of Leeds. He was educated in the Department of Geography, University of Sheffield, and has researched southern African drylands since 1991. His doctoral research focused on the association between environmental change in the Kalahari and changes in soil hydrochemical characteristics. Subsequently, his research in the Kalahari has been extended using satellite remote sensing studies investigating processes of ecological change, further soil chemical investigations, and the integration of social and environmental datasets. He has authored over ten articles on environmental change in the region and the applicability of land degradation assessments.

Robert K. Hitchcock (rhitchco@unlnotes.unl.edu) is professor of Anthropology in the Department of Anthropology and Geography and the Co-ordinator of African Studies at the University of Nebraska-Lincoln. He was educated at the University of California, Santa Barbara, and the University of New Mexico. He has worked in the Kalahari since 1975, primarily on the human ecology of the San and their neighbours and on the impacts of development programmes such as the Tribal Grazing Land Policy, with funding from bodies including the US NSF, Ford Foundation, NORAID, US AID, International Work Group for Indigenous Affairs, and the World Bank. He is the author of numerous articles, chapters, and books on southern African issues, including *Kalahari Communities: Bushmen and the Politics of the Environment in Southern Africa* (Copenhagen: International Work Group for Indigenous Affairs, 1996) and is a co-editor of *Endangered Peoples of Africa and the Middle East: Struggles to Survive and Thrive* (Westport, Conn.: Greenwood, 2001).

Bothepha Kgabung (KGABUNG@mopipi.ub.bw) is a lecturer at the University of Botswana, Department of Environmental Science. She holds BA (majoring in Environmental Science) and M.Sc. (Environmental Science) degrees, both obtained from the University of Botswana. She specializes in natural resource (rangelands and wildlife) management and rangeland ecology. Her areas of research interest include: integrated and sustainable management with focus on community participation and environmental conservation; ecotourism and the relationship between tourism development and wildlife management; and the causes (natural or anthropogenic) and effects of rangeland and wildlife resource dynamics. She has worked in various consultancies and development projects covering EIA's, baseline studies, and investigation of sensitive social, economic, cultural, and environmental issues. The Trans-Kgalagadi Road Project, which traverses the fragile Kalahari ecosystem, is one such project where she was actively involved.

Jeremy S. Perkins (jem@ffmbots.com) is currently co-director of 'Natural Resources and People' (NRP) Pty. Ltd. in Botswana and was previously employed as lecturer in Environmental Science at the University of Botswana. He was educated at the University of Sheffield, where he gained his interest in the Kalahari and from where he conducted his Ph.D. on the impact of borehole dependent cattle grazing on the environment and societies of the eastern Kalahari. He has worked in Botswana and the Kalahari for ten years on a wide range of natural resource management and community based development projects. He has authored numerous articles in academic journals and consultancy reports for government departments and NGOs.

Jacqueline S. Solway (jsolway@trentu.ca) is associate professor of Comparative Development Studies and Anthropology at Trent University, Canada. She has conducted research in Botswana, principally in the Kalahari, for over two decades. She has examined rural socio-economic change, primary health care, and politicized ethnicity. More specifically, she has studied livelihood strategies, patterns of social differentiation, the impact of development policy, and, most recently, the rise of minority group struggle in the context of Botswana's liberal multiparty democracy. Her work has been supported by IDRC and SSHRC and published in books and journals including *Development and Change, Journal of Southern African Studies, Current Anthropology, Ethnos, Canadian Journal of African Studies, Journal of Anthropological Research*.

Deborah Sporton (D.Sporton@sheffield.ac.uk) is lecturer in Human Geography at the University of Sheffield. Her research interests include population geography and the human dimensions of social and environmental change. She has undertaken interdisciplinary research in the Kalahari focused on various aspects of social and environmental change since 1994 funded by the ESRC and DFID with research outcomes appearing in *Land Degradation and Development, Area, Professional Geographer, Journal of Southern African Studies*, and *Applied Geography*.

Greg Stuart-Hill (gstuart@iafrica.com.na) has twenty years of experience in rangeland research and wildlife management in southern Africa. He is currently employed by the World Wildlife Fund (US) as the natural resource adviser for the Community Based Natural Resource Programme in Namibia. He was educated at the University of Natal, South Africa, where he conducted his Ph.D. on the management of the succulent valley bushveld in the eastern Cape. He has worked as: an agricultural researcher, a lecturer at the University of Natal, the head of scientific services for a conservation parastatal, and as an independent ecological consultant. He spent four years in Botswana as team leader to the Botswana Rangeland Inventory and Monitoring Project. He has published in excess of 45 academic papers, and has written a number of chapters in books and numerous consultancy reports.

David S. G. Thomas (d.s.thomas@shef.ac.uk) is a professor of Geography at the University of Sheffield and director of the Sheffield Centre for International Drylands Research. He was educated at the University of Oxford and has conducted research in dryland environments since 1981, principally in the Kalahari region. His research interests include long-term environmental change, dryland geomorphology, and people-environmental interactions, and he is coleader of IGCP project 413: 'Understanding future dryland change from past dynamics'. His research has been funded by bodies including the NERC, ESRC, DFID UK, NATO, UNESCO, and the Royal Society. He is the author of numerous papers and has co-authored and co-edited several books including *The Kalahari Environment* (with Paul Shaw (1991), Cambridge: Cambridge University Press) and *Desertification: Exploding the Myth* (with Nick Middleton (1994), Chichester: Wiley).

Chasca Twyman (c.tywman@sheffield.ac.uk) is a lecturer in Geography at the University of Sheffield. She gained her Ph.D. at Sheffield in 1997, for research into 'Community Development and Wildlife Management: Opportunity and Diversity in Kalahari Wildlife Management Areas, Botswana'. Between 1998 and 2001 she was a Post-Doctoral Research Associate on a Department of International Development funded project investigating poverty, policies, and natural resource use in southern Africa. She has authored a number of papers on her research interests of society–environment interactions, community-based natural resource management, and policy and rural development, principally in southern Africa.

1

Sustainable Livelihoods in Kalahari Environments: A Contribution to Global Debates

David S. G. Thomas and Deborah Sporton

Introduction

The role of the environment now figures prominently in development rhetoric and discourse (World Bank 1992). Peet and Watts (1996) have argued, for example, that poverty and its reduction through the creation of environmentally sustainable livelihoods is central to the environment-development dialectic. Contemporary academic debates informing our understanding of this nexus have shifted in recent years away from normative First World modernist explanations that seek to generalize, to approaches that recognize both the complexity and spatio-temporal diversity of society–environment links. Emphasis is now placed on grass-roots participation and the harnessing of local knowledge to improve livelihoods that are environmentally sustainable and appropriate for successful development.

This collection of essays contributes to these recent debates by examining sustainable livelihoods in the context of the Kalahari environment of southern Africa, a region that is subject to marked spatial and temporal natural variability, and the bounds of which transgress different national policy domains. As such, these essays examine, from a range of disciplinary perspectives, the implications of different policies for rural livelihoods and coping strategies in marginal dryland environments. A number of the contributions are refereed versions of papers presented at the ESRC (Economic and Social Research Council) Global Environmental Change workshop on *Sustainable Livelihoods in Marginal African Environments*, convened by the editors in April 1997. These report research funded by the ESRC, most notably the project on *Environmental Change, Entitlements and Poverty in Kalahari Pastoral Systems*. While the empirical focus of the book is the Kalahari environment, this introductory chapter outlines the wider global context, including the geopolitical, environmental, developmental, and social

frameworks within which enviro-social relations in the Kalahari are structured. The concluding chapter draws together common themes and identifies
enabling and constraining policies and practices that have wider implications,
beyond the Kalahari, for sustainable livelihoods in dryland environments.

Sustainable Development and Livelihood Discourses

It is widely agreed (e.g. Ganry and Campbell 1995; Sen 2000; Markandya
2001) that the concept of sustainable development has its origins in the 1980
World Conservation Strategy Report, but owes its widespread acceptance to
the publicity it received through the Brundtland Report (World Commission
on Economic Development 1987). The report was founded on concerns that
development to date had largely been unsustainable. World development, particularly during the nineteenth and first half of the twentieth centuries, had
been achieved at a cost to future generations. Opportunities for prospering
and survival had been compromised by past actions that affected both the
natural resource base available to indigenous populations and limited the
range of livelihood options open to them. Such actions had often been
grounded, though not exclusively so, in the legacies of the colonial era where
the powerful (the developed countries, or countries of the 'North') had
gained their developmental success at the expense of the exploited (the underdeveloped countries of the 'South'). Rather than reducing the inequalities
that had grown between the North and the South, many development interventions during the immediate post-colonial era may have furthered inequalities and wealth and poverty differentials, while at the same time furthering
activities that are ultimately unsustainable (Long and Long 1992). Here we
briefly consider the central issues underlying debates surrounding the development–environment nexus focusing first on unsustainable development
and natural resource use before exploring key developments towards understanding enviro-social relations following the rise of sustainable development
agendas in the post-Brundtland era.

Unsustainable Development: Colonial Legacies and
Normative Interventions

Global economic development during the eighteenth, nineteenth and early
twentieth centuries was closely related to the processes of colonialism and
the inequalities that it eschewed both within, but principally between, the
colonial power and the colony (Nafziger 1988). Both social and environmental ramifications resulted from the processes associated with this relationship. On the one hand, people in the colonies were exploited, either
directly or indirectly. Direct social impacts included the utilization of local

people as cheap labour by those conducting the development process, which in its most extreme form included slavery and the removal of people from Africa to provide cheap labour in the developing agricultural activities in the Americas. Indirect social impacts included the exclusion of indigenous groups from specific land areas that they may have utilized for generations, in order that the resources of that area could be exploited by the colonial, usually European, group (Ghai and Radwan 1983, Kennedy 1988). This was often achieved through the replacement of, or addition to, customary land tenure systems and institutions by statutory tenure backed up with colonial legal structures (IIED 1999). On the other hand, the process of development had marked environmental impacts, either directly through the utilization, and sometimes exhaustion, of particular natural resources, or indirectly through the by-products of the development process. The social and environmental elements of such development activities have not been isolated processes, however. There are many examples available of instances where the exclusion of people from one area during the colonial period or as a result of large-scale development programmes has increased the pressures on natural resources in another (e.g. Swift 1977 in Somalia, Mortimore *et al.* 1995 in Kenya).

Attempts to move away from the social and environmental legacies of the colonial period in the South have frequently involved the concept of development intervention. During the 1950s, 1960s, and 1970s in particular, these interventions were not always as neutral as they may have initially appeared. Development interventions frequently embraced the normative standards of the North (Nafziger 1988; Hausler 1995), such that they reinforced the processes of 'Westernization, centralization, normalization and standardization of modes of production, forms of government and other organisations . . . [aiming] at the control of patterns of local economic and political development' . . . (Hausler 1995: p. 181). Development often focused upon bringing underdeveloped countries into the global economic system on the terms of the Western world. As a consequence interventions in many instances have resulted in economic benefits that were equal or even greater for the donor country than the recipient (Hausler 1995). They may also have benefited the rich more than the poor in the recipient countries (Blench 1985), thereby increasing inequalities (Markandya 2001). They have in many instances reinforced Western-style power relationships, often centralizing power and leading to the breakdown of local institutions (De Janvry and Garcia 1988), and have continued to displace people from the land (Hitchcock 1980).

Links Between Development, Poverty, and Environmental Degradation

The promotion of rural development during the post-colonial period was often based on the large-scale, often export-led, commercialization of rural production. Founded upon pro-growth principles derived in the North, favouring for example, the privatization of land, such activities are now seen as major contributors to the continued rise in rural poverty and environmental degradation in the countries of the South (Hancock 1989; Lopez 1992; Kipuri 1995; Heath and Binswanger 1996). The displacement of people, to make way for commercialization, deprives them of access to the natural resources that support their livelihoods, which in turn may increase pressures on the areas to which they move, and increase levels of poverty (Sporton et al. 1999).

Markandya (2001) has recently reviewed some of the major people–environment relationships that are often viewed as outcomes of this process. In considering the issue of whether increased poverty leads to increased environmental degradation, it is noted that while the two often do occur together, causality and directionality (poverty causing degradation or degradation causing poverty) is often unproved. There are numerous documented cases, for example, where population displacement to either environmentally marginal areas or densely populated areas has appeared to lead to both increased levels of poverty and the occurrence of environmental degradation (e.g. Tiffen et al. 1994 on the impact of crowding in communal areas in Kenya during the colonial period). The second relationship considered by Markandya is whether environments occupied by the poor are more degraded than those occupied by the wealthy. No clear relationship is identified, but importantly it is noted that in situations where the poor are less able to invest cash in soil conservation strategies, they do substitute other forms of capital at their disposal, notably labour. The interchangeability of different forms of capital in sustainability debates is an issue that has been raised by Sen (2000) and which will be considered later. A third relationship considered by Markandya (2001) is whether a decline in the quality of the environment adversely affects the poor more than the wealthy. It is argued that most of the available evidence supports the assumption that it does, because the poor frequently occupy more marginal lands. This may itself sometimes be a function of historical political economy, for example of colonial actions (Ramphele 1991) and post-colonial development processes. The level to which different groups are affected by environmental degradation will, however, at least in part, be a function of how closely they are tied to the land, and whether other livelihood opportunities exist and can be facilitated (Ellis 1999).

Policy changes, notably those associated with development efforts, may contribute markedly to the nature and direction of people–environment relationships, though they are also mitigated through the effect of factors that are external to the policies themselves such as population change (Lopez 1992). A further element that is widely agreed to have an impact on the environment–poverty relationship is the breakdown in traditional land management and tenure institutions that was associated with the processes of commercialization and privatization of the land (Darkoh 1986; Lopez 1992; Hoffman and Ashwell 2001), a process that has had a close association with the exclusion of indigenous values and knowledge from many development interventions that occurred in the post-colonial period.

From Sustainable Development to Sustainable Livelihoods

The growing recognition of the deleterious impacts of pro-growth, top-down development policies on poverty, livelihoods, and the environment has precipitated new ways of thinking about the development process and policy implementation. From multilateral organizations such as the World Bank and the United Nations Development Programme (UNDP), to NGOs and national governments, down to grass-roots organizations, 'sustainable development' emerged as the panacea.

The central tenet of the concept of sustainable development is well known: sustainable development is that which meets the needs of the present without compromising those of the future. The concept further embraces other key elements: that sustainable development has a role to play in poverty reduction, and that emphasis is not on resource conservation *per se* but on conservation for each generation to meet its needs. While sustainable development remains the dominant discourse underlying poverty alleviation interventions and natural resource management, approaches towards understanding 'sustainability', and the pathways to achieving this goal, have changed in recent years. In particular, the sustainable livelihoods approach has gained increasing recognition among development practitioners and policy-makers alike as a framework for understanding the multiple and dynamic dimensions of livelihoods (see Carney 1998; DFID 1997; Scoones 1998). In summary: 'Livelihoods comprise the capabilities, assets (including both material and social resources) and activities required for a means of living. A livelihood is sustainable when it can cope with and recover from stresses and shocks and manage to enhance its capabilities and assets both now and in the future, while not undermining the natural resource base' (Chambers and Conway, 1992).

Although top-down policy interventions have been replaced by bottom-up, grass-roots, integrated rural development initiatives, the sustainable livelihoods approach, first highlighted in the Brundtland Report (WCED 1987) has been developed in several key ways. Focusing on assets, capabilities, and activities (rather than deprivation), the approach is community-centred and places much greater emphasis on local priorities, interpretations, and abilities with the aim of building on the existing capacities of the rural poor. The environment is treated not as an add-on as in other approaches, but is mainstreamed in this more holistic approach (Carney 1998). Livelihood resources are constituted from complex and dynamic bundles of capital assets comprising tangible, material, and non-tangible dimensions. Capital assets include natural capital (natural resources), social capital (e.g. networks), and human capital (skills and knowledge), as well as financial capital (e.g. cash, savings, credit/debt). Access to these assets, and therefore people's livelihood activities and capabilities, are affected both by institutional arrangements and historical sociopolitical structures. Key to this approach is the recognition of vulnerability and adaptability to livelihood shocks as central to understanding the nature of poverty.

Scoones (1998) identifies five key elements of sustainable livelihoods. The first is linked to rural productivity—livelihoods are sustainable if they create gainful employment either through subsistence production and/or waged labour in activities that enhance the self-worth of rural populations. The second is linked to poverty reduction—for livelihoods to be sustainable the causes of poverty (both qualitative and quantitative) must be mitigated promoting greater equity in access to capital assets. Enhanced capabilities and well-being (Chambers 1997) constitute the third dimension, combining abilities to both access and mobilize assets with more subjective experiences of well-being (feelings of self-esteem, security, happiness). In addition to these three dimensions, Scoones highlights two further elements that are crucial to the sustainability of livelihoods. The fourth is the resilience of livelihoods to short-term stresses and their ability to recover from longer-term shocks. Some livelihoods are more vulnerable to such shocks than others (for example due to lack of diversification, limited access to social or natural capital), making recovery and hence livelihoods less sustainable than others whose capital asset base is perhaps more diversified, or who can, for example, draw on social networks (social capital). The fifth element is the sustainability of the natural resource base—which is the long-term resilience of the natural environment to stresses and shocks. The depletion of natural resources beyond the capacity of a system to maintain productivity may result in the long-term depletion of stocks to the detriment of livelihoods.

Debates about Environmental Degradation

The environment and environmental sustainability is increasingly mainstreamed within contemporary policy approaches aimed at the creation of sustainable livelihoods and poverty alleviation. The scientific bases of what constitutes sustainability, resilience, and environmental degradation have, however, been challenged in recent years.

During the 1970s research conducted among African pastoral societies by Swift (1977), Livingstone (1977), Glantz (1977), and many others contributed to a growing recognition in the North of the complex but rational behaviour of traditional land users at times of environmental stress caused by events such as droughts. Indeed, such systems of resource use could adapt to the challenges of climatic variability, and had done so for centuries, but the addition of institutional and other politico-economic changes contributed to previously sustainable livelihood activities becoming unsustainable (Kipuri 1995).

Work by Huntley (1982), Breman and de Wit (1983), Homewood and Rodgers (1987) also showed how pastoralists understood their environments and their variable grazing resources. This research was closely linked to other detailed ecological investigations that ultimately contributed to the development of a new disequilibrium paradigm to explain the behaviour of dryland ecosystems (Behnke *et al.* 1993). These systems do not follow the succession and climax pathway of temperate ecosystems, but are adapted to naturally varying environmental conditions. They respond to adverse environmental conditions in complex and often multidirectional ways, but ultimately are often highly resilient and have the capacity to recover following the cessation of disadvantageous conditions, for example at the conclusion of a drought period. Consequently, changes in these ecosystems do not necessarily constitute degradation, but can include elements of natural variability.

Environmental degradation has previously been considered in the context of poverty and unsustainable development. But what if environmental degradation has been misconstrued, and wrongly attributed as an expected outcome of indigenous land-use systems? This has occurred, with regard to desertification (Thomas and Middleton 1994) and in respect to the environmental impacts of traditional dryland pastoralism (Mace 1991). Views that the blame for degradation rests with the failure of local institutions and practices, however, have been widely questioned (Turner 1999). Desertification is simply land degradation in drylands (Middleton and Thomas 1997), but has been afforded a special place amongst major global environmental issues due to its association with the Sahel drought of the 1970s and the major United Nations conference concerning desertification held in 1977. It continues to receive significant attention, and is flagged up as a major global environmental issue in Agenda 21, which emanated from UNCED, or the Rio Summit, in 1992. In 1994 the UN general assembly approved an international Convention to Combat Desertification, which has now been ratified by almost two hundred countries, from the North and South, including those that embrace the Kalahari.

Desertification has been highly controversial, with views concerning its representation, characterization, and extent differing markedly (contrast for example Thomas and Middleton 1994 and Stiles 1995). Three particular issues are relevant to this discussion. First, United Nations estimates of its occurrence have been dramatically revised down since 1977. This is due partly to real difficulties in monitoring the occurrence of the range of forms of land degradation that contribute to the overall phenomenon, some of which are subtle. Second, the status of vegetation change as a component of dryland degradation is problematic. This is because of its highly variable nature in time and space and the need to distinguish changes that are an inherent function of natural processes from those that result from human actions on the land and human-induced enhanced climatic changes. Even when this is possible, for example by using remote sensing techniques (Tucker *et al.* 1991) their permanency and significance to human land-use strategies are often contested. Third, the role of society both in contributing to the processes of desertification and in the effective reduction of land degradation were largely ignored and at best underestimated during the 1980s. It has been argued that this contributed to the ineffectiveness of international efforts to reduce or reverse the problem (Maukonen 1995). In contrast, recommendations enshrined within the UN Convention to Combat Desertification (CCD) represent a radical departure from previous approaches to tackling dryland degradation embracing recognition of the links between degradation, poverty, and sustainable development (Thomas 2001). Great importance is attached to the empowerment of local communities and to the role of indigenous knowledge in actions to combat desertification. However, while the CCD espouses many aspects of currently favoured thinking about the relationships between poverty, degradation, and societies, it has also been suggested that its simplicity could hinder a real understanding of the complexities that these relationships embrace (Scoones and Toulmin 1999).

Conventional Wisdom and Indigenous Knowledge

A key problem with previous rural development interventions in the South has been the pre-eminence given to 'conventional' or 'received' (Western) wisdom about the environment, its resilience, and its degradation (which we have shown to be contested) at the expense of indigenous knowledge systems and experience. There are many examples of the imposition of land-use strategies, whether in the forestry, arable, or livestock sectors (Thomas and Middleton 1994), based on Western knowledge and environmental wisdom, often gained in temperate environments where ecosystems function markedly differently from those of the actual environment concerned. In some instances, traditional land-use practices, derived through the experiences of living with the land over many generations, became illegal (Homewood and Rogers 1987; Hausler 1995), as statutory tenure supplanted customary rights. When environmental degradation occurred, it was seen as a result of environmental factors, with little consideration given to the effect of social and political structures. When poverty was prevalent, it was frequently viewed not as a function of the development process but of the lack of integration of rural peoples in the market economy that was being developed (Hausler 1995). Population growth was widely seen as a contribution to both degradation and poverty, though its precise role has been controversial and widely debated (Boserup 1993; Markandya 2001). It has been suggested that in some instances, with appropriate institutional frameworks, poverty reduction and reduced environmental degradation can be achieved despite high population densities (Tiffen *et al.* 1994).

Since the 1980s the principles underlying 'traditional' development policy interventions have been increasingly questioned (Bollig and Schulte 1999). This has been associated with their perceived failure to improve the well-being of people in the South and with growing concerns about the state of the environment, suggesting that many practices that had arisen in the colonial and post-colonial periods were socially and environmentally unsustainable. Linked closely with this process has been a paradigm shift amongst external 'experts' regarding the value of indigenous knowledge and traditional resource-use practices (Leach and Mearns 1996). These practices are no longer viewed as economically irrational (Livingstone 1977) and do not necessarily cause environmental degradation but may achieve the opposite (Livingstone 1991). It is now recognized that external interventions and the introduction of alien, Western approaches to environmental management have not enhanced well-being (Kipuri 1995; Barrow 2000). Above all, traditional indigenous systems of natural capital use are now increasingly viewed by external agencies not as unsustainable and irrational, but as sustainable and highly logical.

Enviro-development Issues in the Kalahari

Many of the central issues explored above are relevant to understanding both developmental and environmental issues that have been affecting the Kalahari of southern Africa. European interventions in the Kalahari region during the nineteenth century largely involved hunting wildlife and trading the resultant products for goods manufactured in Europe. European traders increasingly gained a foothold in the region during the nineteenth century, but patterns and routes of trade tended to follow those that had developed since the iron age (Denbow and Wilmsen 1986). Livestock increasingly entered the Kalahari during the nineteenth century but on a transhumance basis and under the ownership of African, principally Tswana and Herero, rather than European, pastoralists. By the end of the nineteenth century the combined effects of drought, hunting, and disease had depressed wildlife and livestock numbers significantly. Moreover, the Kalahari environment was left degraded, and characterized, in some areas, by the replacement of grass species with bush cover (Campbell and Child 1971) such that the region was regarded as a remote backwater of the British Empire (Thomas and Shaw 1991). An exception to this condition applied to some peripheral areas, notably the northern Cape of South Africa, the Molopo valley (on the border between present-day South Africa and Botswana), and the Ghanzi Ridge (north-west Botswana), where farm blocks were demarcated for European settlers in the 1890s. The Ghanzi ridge in particular was a relatively well-watered area by dint of its geological structure, such that commercial farms were established there depriving hunter-gatherer and pastoralist groups of access to water resources.

Pastoralism slowly expanded during the twentieth century through the sinking of boreholes during the 1930s to increase grazing opportunities and to provide relief during times of drought in Bechuanaland. Veterinary controls, to limit the outbreak and spread of bovine diseases, commenced in 1907 (Falconer 1971), and the first veterinary cordon fences were constructed in the Kalahari in 1954 (Thomas and Shaw 1991). In northern Botswana various attempts were made to increase the range of livestock through the control of the tsetse fly. Despite increasing European interest in the cattle industry, livestock production in the Kalahari remained a low-intensity affair throughout the colonial period of the twentieth century, and in the case of Bechuanaland was principally concentrated in the wetter areas to the east of the Protectorate. Two actions during this period are likely to have had a marked influence of the post-independence growth of the industry in the Kalahari. First, Yeager (1989) has noted how the increasing colonial interest in the cattle industry did impart a significant degree of commercialization on Tswana herders. Second, the sinking of boreholes in the Kalahari increased markedly from the 1950s onwards, first as a drought relief measure but increasingly as a contribution to the expansion of the cattle industry (Mazonde 1988) into areas regarded as highly suitable for grazing (Debenham 1952).

Following the independence of Botswana in 1966, the nature of development and natural resource use in the Kalahari has been influenced markedly by two changes that have exerted a major influence. Changes in land allocation procedures, particularly the emergence of land boards, transferred responsibility for the allocation and distribution of land away from customary institutions. This was accompanied by a drive to increase cattle production, supported through preferential access to European markets, which led to the growth of formal year-round livestock production which was increasingly focused on leasehold ranches in the Kalahari.

Much has been written about the impact in particular of the Tribal Grazing Lands Policy (TGLP), the third large-scale livestock policy that was implemented after 1970, on the societies, environment, and development of the Kalahari (e.g. Hitchcock 1978; Odell 1980; Perkins and Thomas 1993; Perkins 1996). While Hitchcock (1978) considered the impact of the borehole-centred TGLP ranches and other Kalahari cattleposts on the livelihoods of mobile Khoisan groups in the Kalahari, many of the early assessments of their effects was focused on their environmental impacts. Williamson and Williamson (1985) and others focused upon the effect of TGLP perimeter fences and the associated increase in the trans-Kalahari network of veterinary control fences on wildlife populations. The effects on vegetation communities and soil quality of year-round livestock pressures were investigated by Perkins and Thomas (1993) and others. The impact of year-round livestock production in the Kalahari has sometimes been viewed as leading to rapid and significant desertification (e.g. Ringrose and Mathieson 1987). More recently, the nature of ecosystem changes and whether they really do represent degradation have been contested (e.g. White 1993) in the light of developments in ecological theories (Dougill *et al.* 1999). Arguably, degradation is most relevant as a concept when considered in terms of its impact on livelihood opportunities and land management practices (e.g. Abel and Blaikie 1989): it has to have a human dimension to be relevant. A survey in the northern Kalahari by Chanda (1996) assessed perceptions of degradation and its causes and solutions among rural communities. The study revealed a high level of recognition of degradation issues among those consulted, although they were largely fatalistic about its causes and solutions, viewing their own contributions as limited. Van Rooyan (1998) noted a similar attribution of causality amongst the residents of the south-west Kalahari in the Northern Cape, South Africa.

Van Rooyan (1998) suggests that environmental degradation can be tackled only if the land users themselves contribute centrally to the identification of the problems involved and if their solutions are embodied within the overall livelihood goals and options that are available. It is significant that the environmental changes associated with fencing and livestock production in the core of the Kalahari are now being considered in the wider context of interactions between levels of well-being, access to resources, and degradation. This is the perspective taken by the contributors to this book, who demonstrate the complex relationships between Kalahari residents, their livelihoods, and their environment.

Format of this Book

The nine main chapters comprise essays each of which makes a contribution to understanding the issues and pressures surrounding sustainable livelihoods in the Kalahari region. The essays do not attempt to provide total thematic coverage of the issues involved, nor do they attempt to provide complete spatial coverage across the whole of the Kalahari region. Rather, each provides vital insights into major themes that are central to issues that affect sustainability and livelihoods for the people of the Kalahari. In each chapter one or more of the key tenets of environment, policy, and structural change provides the central element around which the sustainable livelihoods theme is considered. Box 1.1, at the end of this chapter, provides a summary of the different ethnic groups encountered within the Kalahari and which appear in the various chapters of the book.

In Ch. 2 the Kalahari environment, the 'natural' element that underpins resource-based livelihoods, is considered from the perspective of its salient

Box 1.1. Summary of key ethnic groups of the Kalahari region

1. Basarwa

Also referred to in the literature as San, Sarwa, Bushmen, N/oakwe, or simply 'the first people'. Indigenous former foraging peoples occupying the Kalahari and adjacent areas, taken into serfdom by the Bakgalagadi and Batswana during the nineteenth century. Today they predominate amongst Remote Area Dweller (RAD) groups and also work with live-stock as herders supplemented by gathering and hunting. The main Basarwa groups: Tsasi, Naro, G/wi, !Xo, Balala, Kua (Lee and DeVore 1968; Lee 1979; Hitchcock 1978; Barnard 1992).

2. Bakgalagadi

Early Sotho-Tswana settlers on fringes and within the Kalahari com-prising several groups. Four main groups: Kgwatheng, Boloengwe, Ngologa, Shaga (Hitchcock and Campbell 1980; Kuper 1969, 1970; Solway 1980). Also includes the Phaleng and Pedi in Central District (Hitchcock and Campbell 1980). Traditionally they were mobile with cattle herds practising hunting and gathering, now largely sedentarized.

3. Batswana

The main population group in Botswana, subdivided into numerous tribes. They arrived in the region in a series of migrations between the seventeenth and early nineteenth centuries, pushing earlier inhabitants such as the Basarwa and Bakgalagadi westwards into the Sandveld. Today they are principally sedentary agro-pastoralists occupying the wetter eastern areas of Botswana. Main Batswana groups: Bangwato, Batawana, Bakhurutse, Bakwena, and Bakaa (Schapera 1930, 1953).

4. Non-Tswana Sedentary Groups

This category refers mainly to the Herero, Bantu-speakers who arrived in Botswana as destitute refugees from the Herero-German war in Namibia (1904–7). Many of those who survived the German genocide fled into Botswana, to Ghanzi, Ngamiland, and Central Districts (Vivelo 1977). Others fled to Kgalagadi District (Jerve 1985). They are princi-pally cattle farmers who raise a few crops.

characteristics and how these influence livelihood opportunities. The Kalahari is often presented as a homogenous dry savanna ecosystem, however, it is demonstrated here that spatial and temporal heterogeneity are in fact key facets of the vital resource elements of water (including drought), vegetation, and wildlife. These elements have been keys to successful subsistence livelihoods, but it is shown how, with the onset of progressive expansion of commercial utilisation of the environment during the twentieth century, heterogeneity has both confounded and structured management attempts. The Kalahari's heterogeneity is not static however, and the chapter considers important aspects of natural and human-induced variability in the availability of, and access to, water resources, plant community composition, and animals. Four important issues or challenges are also considered: management of the environment; sedentary versus mobile activities; the distinction between natural environmental variability and environmental degradation; the distinction between global warming and sustainable resource use. Selected themes in this chapter are developed in more detail in Chs. 4 and 5.

The historical context and political and social frameworks in which envirosocial relations are embedded and discursively produced in the Kalahari (comprising its core country, Botswana, and neighbouring Namibia and South Africa) are discussed in Ch. 3 with particular reference to land assets. The development of a largely unregulated cattle industry in Bechuanaland prior to independence as Botswana in 1966 is contrasted with the policy-driven displacement of indigenous groups in the Northern Cape, South Africa, and eastern South-West Africa during the colonial and apartheid eras. Within each of these policy domains the challenge has been to respond to the legacy of colonial and subsequent apartheid land policies which have created an underclass of rural poor, deprived of the capital assets necessary to sustain livelihoods. Land restitution, resettlement, and settlement policies in post-independence times, including the debates and dilemmas surrounding them, are discussed for the Kalahari regions of South Africa, Namibia, and Botswana respectively, illustrating the complex structural issues that are arising. In particular conflicts between access to land and the continued commercialization and hence preservation of the racialized status quo are highlighted, which are associated with some of the key environmental issues considered in the previous chapter. The constraints and opportunities for livelihoods and natural resource use presented by structural changes are examined as a precursor to considering their possible impacts on sustainable opportunities for Kalahari populations.

Chapters 4 to 7 consider key facets of the Kalahari land-use changes—particularly commercialism of the livestock industry, fencing, and the creation of specific areas in which traditional livelihood activities are encouraged—and their implications for sustainable rural livelihoods and sustainable land use. In Ch. 4 Jeremy Perkins and co-workers consider the environmental and resource implications of one of the primary facets of the commercialization of the Kalahari: fencing. Fencing began in earnest in the core of the Kalahari at Botswana's independence, and it remains the thrust of the New Agricultural Policy. Fencing has been seen as the panacea to various ills in the Kalahari, notably disease and problem animal control, but has been carried out despite running against some core ecological principles and despite its impact on communal lands, access to water, and wildlife populations. The authors argue that the resource-use problems that fencing is attempting to resolve are misconceived, at a cost to rural livelihood, and wildlife, opportunities.

In Ch. 5 Andrew Dougill considers the environmental implications of vegetation change on the formalized ranches that have been created by fencing. An increase in bush species has been clearly linked to intensified cattle grazing. However, contentious debates remain, revolving around whether these ecological changes are permanent, and therefore represent land degradation, or irreversible, and therefore simply a consequence of sustainable agricultural use. To assess the permanence of ecological changes, this chapter investigates the impact of intensified cattle grazing on the key determinants that affect vegetation growth, namely soil water and nutrient availability patterns.

A key policy associated with the commercialization of livestock production and the creation of fenced ranches, Botswana's Tribal Grazing Land Policy (TGLP), is discussed by Deborah Sporton and David Thomas in Ch. 6. The TGLP aimed to enhance livestock production and ameliorate perceived rangeland degradation, challenged in Ch. 5, while at the same time improving the livelihoods of rural populations who would be provided with alternative, centralized livelihood opportunities at service centres elsewhere. The chapter highlights the variable nature of policy implementation through space and over time and in particular how livelihood opportunities were enhanced under TGLP where policy implementation was participatory, flexible, and adapted to environmental conditions.

A further outcome of the fencing of the Kalahari in Botswana and the demarcation of land areas for specific activities has been the creation of Wildlife Management Areas (WMAs). In Ch. 7 Chasca Twyman examines the changing societal–environmental relationships in the Kalahari arising from the establishment of WMAs in 1986. They are multiple-use areas combining wildlife conservation with the establishment of self-sufficient, sustainable rural economies. The rural populations (known as Remote Areas Dwellers or RADs) are permitted access to a wide range of natural resources (land, plants, water, and wildlife) in the WMAs. Botswana's WMAs are unique: 'traditional' land-use practices, widely documented by anthropologists, have been displaced to these areas in the wake of commercial pastoralist expansion elsewhere. However, scant attention has been focused on the extent to which these displaced livelihoods and activities, lauded by environmentalists as sustainable, have been adapted in WMAs.

Perhaps the principle environmental variable that impinges upon Kalahari residents is the common occurrence of drought, a phenomenon that is embodied in the issues raised in Chs. 8 and 9. In Ch. 8 Robert Hitchcock explores the dual pressures exerted on northern Kalahari rural populations by cycles of drought and livestock disease. The success of government safety-net policy interventions are discussed alongside the ways in which rural people themselves draw upon indigenous environmental knowledge and coping strategies to restructure and diversify livelihoods in response to drought and disease disturbances.

Jacqueline Solway considers in Ch. 9 the wider role for drought in accelerating social change in the Kalahari. Drought is considered to be a 'revelatory crisis' in which structural contradictions as well as deteriorating socio-economic conditions are exposed. Paradoxically, however, drought is also seen as enabling the concealment of these conditions which can be attributed to the 'crisis' and not to deeper problems and trends. In addition, crises such as droughts disrupt conventional routine sufficiently to allow groups including policy-makers and rural dwellers to innovate with normative codes. This fact along with the opening up of structural fault lines often leads to accelerated rates of social change. Change is analysed here both in terms of the structural conditions in which it takes place (and alters) and with regard to the actions taken by individuals and institutions in order to reveal the links between structure and agency. The chapter draws upon an extended case study from Central Botswana and utilizes Sen's entitlement analysis to examine changing social processes.

The final chapter identifies key themes underlying enviro-social relations and sustainable livelihoods in the Kalahari region that resonate across the individual chapters and which inform wider global debates. In particular, historically rooted, racialized systems of social relations underpin access to livelihood assets such that, without redress, policy interventions may provide only short-term solutions, concealing longer-term problems and exacerbating social polarization. Vulnerability and risk are key elements affecting the sustainability of livelihoods. Within the Kalahari, drought and rainfall variability represent a risk to the livelihoods of the most vulnerable and as such have constituted a pervasive theme throughout the chapters of the book. From the perspective of livelihoods, it is not climatic variability *per se* that is the primary problem, but the risks that the variability induces, and people's vulnerability to the effects of these risks, that are more critical. While the focus of debates within environmental science now question the permanence, or irreversibility, of degradation, the significance of local practices to livelihoods that challenge even current environmental orthodoxies are highlighted in several chapters and are explored within the context of environmental management strategies. The policy challenge for the twenty-first century for dryland regions is the resolution of conflicting pressures between external forces favouring the further commercialization and commodification of the environment to the detriment of local livelihoods and those which emphasize the goal of grass-roots empowerment to achieve sustainable rural development.

References

ABEL, N. O. J., and BLAIKIE, P. M. (1989), 'Land degradation, stocking rates and conservation policies in the communal rangelands of Botswana and Zimbabwe', *Land Degradation and Rehabilitation*, 1: 101–23.

BARNARD, A. (1992), *Hunters and Herders of Southern Africa: A Comparative Ethnography of the Khoisan Peoples* (Cambridge: Cambridge University Press).

BARROW, E. (2000), 'People in arid lands: beyond the technical fix', *World Conservation*, 2: 6–8.

BEHNKE, R. H., SCOONES, I., and KERVEN, C. (eds.) (1993), *Range Ecology at Disequilibrium* (London: ODI).

BLENCH, R. (1985), 'Pastoral labour and stock alienation in the sub-humid and arid zones of West Africa', *Pastoral Network Paper 19e* (Cambridge: ODI).

BOLLIG, M., and SCHULTE, A. (1999), 'Environmental change and pastoral perceptions: Degradation and indigenous knowledge in two African pastoral communities', *Human Ecology*, 27: 493–511.

BOSERUP, E. (1993), *The Conditions of Agricultural Growth: The Economics of Agrarian Change under Population Pressure* (London: Earthscan Publications).

BREMAN, H., and DE WIT, C. (1983), 'Rangeland productivity and exploitation in the Sahel', *Science*, 221: 1341–7.

CAMPBELL, A. C., and CHILDS, G. F. (1971), 'The impact of man on the environment of Botswana', *Botswana Notes and Records*, 3: 91–109.

CARNEY, D. (ed.) (1998), *Sustainable Rural Livelihoods: What Contribution Can We Make?* (London: DFID).

CHAMBERS, R. (1997), 'Responsible well-being—a personal agenda for development', *World Development*, 25: 1743–5.

CHAMBERS, R., and CONWAY, G. (1992), 'Sustainable rural livelihoods: practical concepts for the 21st century', *IDS Discussion Paper*, 296 (Brighton: IDS).

CHANDA, R. (1996), 'Human perceptions of environmental degradation in a part of the Kalahari ecosystem', *Geojournal*, 39: 65–71.

DARKOH, M. B. K. (1986), 'Combating desertification in Zimbabwe', *Desertification Control Bulletin*, 13: 17–24.

DEBENHAM, F. (1952), 'The Kalahari today', *Geographical Journal*, 118: 12–23.

DE JANVRY, A., and GARCIA, R. (1988), 'Rural poverty and environmental degradation in Latin America: Causes, effects and alternative solutions', *IFAD paper S 88/1/L3* (Rome: IFAD).

DENBOW, J. R., and WILMSEN, E. N. (1986), 'Advent and course of pastoralism in the Kalahari', *Science*, 234: 1509–15.

DOUGILL, A. J., THOMAS, D. S. G., and HEATHWAITE, A. L. (1999), 'Environmental change in the Kalahari: Integrated land degradation studies for nonequilibrium dryland environments', *Annals, Association of American Geographers*, 89: 420–42.

ELLIS, F. (1999), 'Rural livelihood diversity in developing countries: Evidence and policy implications', *ODI Natural resource Perspectives*, 40 (London: ODI).

FALCONER, J. (1971), 'History of the Botswana Veterinary Services 1905–1966', *Botswana Notes and Records*, 3: 74–8.

GANRY, F., and CAMPBELL, B. (eds.) (1995), *Sustainable Land Management in African Semi-arid and Subhumid regions*, Proceedings of the SCOPE Workshop, November 1993, Dakar, Senegal (Montpellier: CIRAD-CA), 15–19.

GHAI, D., and RADWAN, S. (eds.) (1983), *Agrarian Policies and Rural Poverty in Africa* (Geneva: ILO).

GLANTZ, M. (1977), 'Water and inappropriate technology: Deep wells in the sahel', in V. P. Nanda (ed.), *Water Needs for the Future* (Boulder: Westview), 305–18.

HANCOCK, G. (1989), *Lords of Poverty* (London: Macmillan).

HAUSLER, S. (1995), 'Listening to the people: The use of indigenous knowledge to curb environmental degradation', in D. Stiles (ed.), *Social Aspects of Sustainable Dryland Management* (London: Wiley), 179–88.

HEATH, J., and BINSWANGER, H. (1996), 'Natural resource degradation effects of poverty and population growth are largely policy-induced: The case of Columbia', *Environment and Development Economics*, 1: 65–84.

HITCHCOCK, R. K. (1978), *Kalahari Cattle Posts: A Regional Study of Hunter-Gatherers, Pastoralists and Agriculturalists in the Western Sandveld Region, Central District, Botswana* (Botswana: Government Printer).

——(1980), 'Tradition, social justice and land reform in central Botswana', *Journal of African Law*, 24: 1–34.

HITCHCOCK, R. K., and CAMPBELL, A. (1980), 'Settlement Patterns of the Bakgalagadi', Paper presented at the Symposium on Settlement, Botswana Society, Gaborone.

HOFFMAN, T., and ASHWELL, A. (2001), *Nature Divided: Land Degradation in South Africa* (Cape Town: University of Cape Town Press).

HOMEWOOD, K., and RODGERS, W. A. (1987), 'Pastoralism, conservation and the over-grazing controversy', in D. M. Anderson and A. T. Grove (eds.), *Conservation in Africa: People, Policies and Practice* (Cambridge: Cambridge University Press), 111–28.

HUNTLEY, B. (1982), 'Southern African savanna', in B. Huntley and B. Walker (eds.), *Ecology of Tropical Savanna* (Berlin: Springer-Verlag), 101–19.

IIED (1999), *Land Tenure and Resource Access in West Africa: Issues and Opportunities for the Next Twenty-Five Years* (London: International Institute for Environment and Development).

JERVE, A. M. (1985), *Cattle and Inequality: A Study in Rural Economic Differentiation from Southern Kgalagadi in Botswana* (Bergen: DERAP).

KENNEDY, P. (1988), *African Capitalism: The Struggle for Ascendancy* (Cambridge: Cambridge University Press).

KIPURI, N. (1995), 'Socio-economic concerns on sustainable use and management of semi-arid lands: The case of pastoral lands of East Africa', in F. Ganry and B. Campbell (eds.), *Sustainable Land Management in African Semi-arid and Subhumid*

Regions, Proceedings of the SCOPE Workshop, 15–19 November 1993, Dakar, Senegal (Montpellier: CIRAD-CA), 273–82.

KUPER, A. (1969), 'The Kgalagadi in the Nineteenth Century', *Botswana Notes and Records*, 2.

——(1970), *Kalahari Village Politics* (Cambridge: Cambridge University Press).

LEACH, M., and MEARNS, R. (eds.) (1996), *The Lie of the Land: Challenging Received Wisdom on the African Environment* (Oxford: James Currey).

LEE, R. B. (1979), *!Kung San: Men Women and Work in a Foraging Society* (New York: Cambridge University Press).

LEE, R. B., and DeVORE, I. (eds.) (1968), *Man the Hunter* (Chicago: Aldine).

LIVINGSTONE, I. (1977), 'Economic irrationality among pastoral peoples: Myth or reality?', *Development and Change*, 209–30.

——(1991), 'Livestock management and "overgrazing" among pastoralists', *Ambio*, 20: 80–5.

LONG, N., and LONG, A. (1992), *Battlefields of Knowledge* (London: Routledge).

LOPEZ, R. (1992), 'Environmental degradation and economic openness in LDCs: The poverty linkage', *American Journal of Agricultural Economics*, 74: 1138–45.

MACE, R. (1991), 'Overgrazing overstated', *Nature*, 349: 280–1.

MARKANDYA, A. (2001), *Poverty Alleviation and Sustainable Development: Implications for the Management of Natural Capital*, paper prepared for the Workshop on Poverty and Sustainable Development (Ottawa: International Institute for Sustainable Development), 21 p.

MAUKONEN, T. (1995), 'Sustainable dryland development and desertification control', in F. Ganry and B. Campbell (eds.), *Sustainable Land Management in African Semi-arid and Subhumid Regions*, Proceedings of the SCOPE Workshop, 15–19 November 1993, Dakar, Senegal (Montpellier: CIRAD-CA), 15–18.

MAZONDE, I. N. (1988), 'The inter-relationship between cattle and politics in Botswana's economy', in G. J. Stone (ed.), *The Exploitation of Animals in Africa* (Aberdeen: Aberdeen University African Studies Group), 345–56.

MIDDLETON, N. J., and THOMAS, D. S. G. (eds.) (1997), *World Atlas of Desertification*, 2nd edn. (London: UNEP/Edward Arnold).

MORTIMORE, M., TIFFEN, M., and GICHUKI, F. (1995), 'Sustainable growth in Machakos, Kenya', in D. Stiles (ed.), *Social Aspects of Sustainable Dryland Management* (London: Wiley), 131–43.

NAFZIGER, E. W. (1988), *Inequality in Africa: Political Elites, Proletariat, Peasants and the Poor* (Cambridge: Cambridge University Press).

ODELL, M. L. (1980), 'Botswana's first livestock development project—an experiment in agricultural transformation', unpublished report (Gaborone: Swedish International Development Agency).

PEET, R., and WATTS, M. (eds.) (1996), *Liberation Ecologies. Environment, Development, Social Movements* (London: Routledge).

PERKINS, J. S. (1996), 'Botswana: Fencing out the equity issue. Cattle and cattle ranches in the Kalahari Desert', *Journal of Arid Environments*, 33: 503–17.

PERKINS, J. S., and THOMAS, D. S. G. (1993), 'Spreading deserts or spatially confined environmental impacts? Land degradation and cattle ranching in the Kalahari of Botswana', *Land Degradation and Rehabilitation*, 4: 179–94.

RAMPHELE, M. (1991), 'New day rising', in M. Ramphele and C. McDowell (eds.), *Restoring the Land: Environment and Change in Post-Apartheid South Africa* (Johannesburg: The South), 1–12.

RINGROSE, S., and MATHIESON, W. (1987), 'Desertification in Botswana: Progress towards a viable monitoring system', *Desertification Control Bulletin*, 13: 6–11.

SCHAPERA, I. (1930), *The Khoisan Peoples of Southern Africa* (London: Routledge & Kegan Paul).

——(1953), *The Tswana* (London: International African Institute).

SCOONES, I. (1998), 'Sustainable rural livelihoods. A framework for analysis', *IDS Working Paper 72* (Brighton: IDS).

SCOONES, I., and TOULMIN, C. (1999), *Policies for Soil Fertility Management in Africa* (London: IIED).

SEN, A. (2000), 'The ends and means of sustainability', keynote address, International Conference on Transition and Sustainability.

SOLWAY, J. S. (1980), 'People, cattle and drought in the western Kweneng District', *Rural Sociology Report Series*, 16 (Gaborone: Ministry of Agriculture).

SPORTON, D., THOMAS, D. S. G., and MORRISON, J. (1999), 'Outcomes of social and environmental change in the Kalahari of Botswana: The role of migration', *Journal of Southern African Studies*, 25: 441–59.

STILES, D. (1995), 'An overview of desertification as dryland degradation', in D. Stiles (ed.), *Social Aspects of Sustainable Dryland Management* (Chichester: Wiley), 3–20.

SWIFT, J. (1977), 'Pastoral development in Somalia: Herding cooperatives as a strategy against desertification and famine', in M. H. Glantz (ed.), *Desertification. Environmental Degradation in and around Arid Lands* (Boulder, Col.: Westview), 275–305.

THOMAS, D. S. G. (2001), 'Into the third millennium: The role of stakeholder groups in reducing desertification', in A. S. Alsharhan, W. W. Wood, and A. S. Goudie (eds.), *Desertification in the Third Millennium* (Rotterdam: Balkema).

THOMAS, D. S. G., and MIDDLETON, N. J. (1994), *Desertification: Exploding the Myth* (Chichester: Wiley).

THOMAS, D. S. G., and SHAW, P. A. (1991), *The Kalahari Environment* (Cambridge: Cambridge University Press).

TIFFEN, M., MORTIMORE, M., and GICHUKI, F. (1994), *More People, Less Erosion: Environmental Recovery in Kenya* (Chichester: Wiley).

TUCKER, C. J., DREGNE, H. E., and NEWCOMB, W. W. (1991), 'Expansion and contraction of the Sahara Desert from 1980 to 1990', *Science*, 253: 299.

TURNER, M. D. (1999), 'Conflict, environmental change and social institutions in dryland Africa: Limitations of the community resource management approach', *Society and Natural Resources*, 12: 643–57.

VAN ROOYAN, A. F. (1998), 'Combating desertification in the southern Kalahari: Connecting science with community action in South Africa', *Journal of Arid Environments*, 39: 285–97.

VIVELO, F. R. (1977), *The Herero of Western Botswana* (New York: West Publishing).

WHITE, R. (1993), *Livestock Development and Pastoral Production on Communal Rangeland in Botswana* (Gaborone: Botswana Society).

WILLIAMSON, D. T., and WILLIAMSON, J. E. (1985), 'Botswana's fences and the depletion of Kalahari wildlife', *Parks*, 10: 5–7.

WORLD BANK (1992), *Development and the Environment*, World Development Report (Washington: World Bank).

WORLD COMMISSION on ECONOMIC DEVELOPMENT (1987), *Our Common Future*, Report of the World Commission on Environment and Development (Oxford: Oxford University Press).

YEAGER, R. (1989), 'Demographic pluralism and ecological crisis in Botswana', *Journal of Developing Areas*, 23: 385–404.

2

Sand, Grass, Thorns, and . . . Cattle: The Modern Kalahari Environment

David S. G. Thomas

The Kalahari Defined

The extensive elevated, flat, sand-covered plain that occupies a substantial part of the southern African interior has variously been described as a desert, thirstland, or savanna. All are appropriate, for this area, known as the Kalahari, is in fact more heterogeneous than it has often been portrayed. The most generous definitions refer to an area of some 2.5 million km^2 that extends from the vineyards on the margins of the Orange River at Upington, South Africa (latitude 29°S) to north of the Congo River into the south-east corner of equatorial Gabon. This area gains its unity from the prevalence of aeolian-origin Kalahari sand as the dominant surface sediment (Thomas and Shaw 1991), thus giving it a geological definition.

At the other extreme, the Kalahari is sometimes described as the area comprising the south-western corner of Botswana, to the west of Tsabong, and adjacent parts of the Northern Cape Province of South Africa; an area that is climatically arid and desert-like in appearance, with episodically active sand dunes. This is the Kalahari Dune Desert (Thomas 1986), the driest part of the interior with an average rainfall of less than 150mm per annum and a high drought incidence. This area, which is relatively small and covers less than 80,000 km^2, also represents the popular image of the Kalahari. Sparsely grassed sand dunes, extending in parallel fashion to the horizon, dry watercourses dotted with occasional acacia trees. Impala, gemsbok, meerkats, and lions (all present, in large numbers, in the Kalahari Transfrontier Park, centred on the dry Nossop and Aoub valleys), and khoisan 'bushmen' (less common) complete the popular representation of the Kalahari. All these images are represented in the marketing of the recently launched Kalahari Thirstland Liquor—a South Africa produced beverage 'made from bulbs, plants and herbs of southern Africa' and sold at airport duty-free stores in the region in pottery imitation gourds.

Between the two lies the Kalahari as usually presented in scientific, political, and administrative usage. An area, coincident with the western three-quarters of Botswana, eastern Namibia, and the Northern Cape Province, extending from the Orange River north to the Caprivi Strip at *c*.18°S, it embraces ecosystems as diverse as the dune desert and the Okavango Delta. This Kalahari is dominated by savanna vegetation systems, which have been increasingly modified by human actions. The region is largely devoid of permanent surface water except in the south with the Orange River and in the north with the Okavango, Cunene, and Chobe rivers, all with their origins in tropical highland to the north in Angola. With the exception of a few towns in peripheral areas, such as Maun, Kasane, and Ghanzi in Botswana, Gobabis in Namibia, and Upington in South Africa, and self-contained mineral (notably diamond) extraction operations that are locally significant but spatially small and self-contained, the Kalahari has a predominantly rural economy based upon pastoral activities, which are both longstanding and subsistent, and more recent and commercial. This is the Kalahari that this book focuses upon; a sandy, largely semi-arid region that none the less contains within it notable environmental contrasts and which today, with the pressures of development agendas and social and environmental change, presents considerable challenges to its residents and administrators. This chapter sets out to characterize and explain the nature of the Kalahari environment. There are already a number of analyses of the Kalahari environment that have been published (e.g. Arntzen and Veenendaal 1986; Thomas and Shaw 1991) such that it is not necessary to repeat them, and to which the reader is referred for greater detail. This chapter will provide a summary of the salient points that are relevant to debates about sustainable livelihoods in the region, outlining the major characteristics of the Kalahari and challenges for its sustainable use.

Kalahari Characteristics

Water: Rainfall, Surface Water, and Groundwater

The Kalahari lies within the southern African summer rainfall zone. The location of the inter-tropical convergence zone over or close to the region in the southern hemisphere summer months draws in moisture-bearing air masses and enhances atmospheric instability. Most of the moisture drawn to the region is derived from the Indian Ocean, so that precipitation, which usually falls during afternoon and evening storms, decreases in a south-westerly direction, both in total amounts and the duration of the wet season. Maun, on the southern edge of the Okavango Delta in central northern Botswana, has a mean annual rainfall of over 500 mm, Tsabong, the largest town in south-western Botswana, receives on average less than 200 mm each

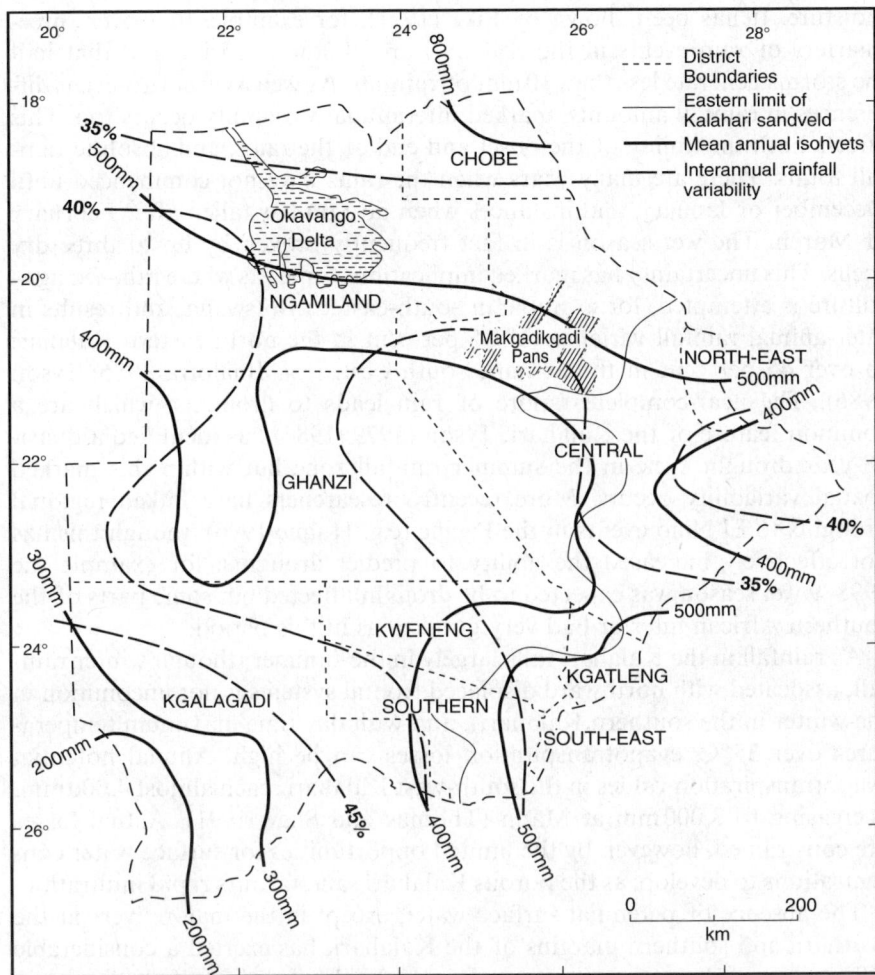

Fig. 2.1. Mean annual rainfall and interannual variability in the Kalahari.

year, while the 'Meir' country of the Northern Cape, west of the Nossop valley, may have a mean value of less than 150 mm per annum (Fig. 2.1). The wet season usually commences within the months of October to November, getting later towards the south-west, where it tends to finish in early April, stretching to May in north-eastern locations.

This description ignores, however, two of the principal characteristics of Kalahari rainfall: variability and uncertainty. Although Kalahari rainfall is associated with convection, it is important to note that this does not mean that each rain event is heavy or that they each bring significant amounts of moisture. It has been shown by Pike (1971), for example, that over three-quarters of rain events in the Kalahari are of low intensity, and that half the storms generate less than 10 mm of rainfall. As well as event-to-event differences in rainfall amounts, marked interannual variability occurs too. This affects both the timing of the onset and end of the rains, and absolute rainfall totals. There are many years when the rains have not commenced until December or January, and instances when no rain has fallen after February or March. The wet season is in fact frequently broken by up to three dry spells. This uncertainty has marked implications for areas where rain-fed agriculture is attempted, for example in south-eastern Botswana, and results in inter-annual rainfall variation of 35 per cent in the north-eastern Kalahari to over 45 per cent in the extreme south-west, (e.g. Bhalotra 1985; Tyson 1986). The near-complete failure of rain leads to droughts, which are a common feature of the Kalahari. Tyson (1979, 1986) has identified a quasi-18-year drought cycle in the summer rainfall zone but within this marked spatial variability occurs. More recently, researchers have linked regional droughts to El Niño events in the Pacific (e.g. Hulme 1996), though this has not effectively increased the ability to predict droughts; for example the 1998–9 wet season was expected to be drought affected but some parts of the southern African interior had very good rains in this period.

As rainfall in the Kalahari falls largely in the summer (though winter rainfall, assocated with northward displaced frontal systems is not uncommon in the winter in the southern Kalahari), and with day time maximum temperatures over 35 °C, evapotranspiration losses can be high. Annual potential evapotranspiration values in the south-west Kalahari reach almost 4,000 mm, decreasing to 3,000 mm at Maun (Thomas and Shaw 1991). Actual losses are constrained, however, by the limited opportunities for surface water concentrations to develop, as the porous Kalahari sand favours rapid infiltration.

The absence of perennial surface water, except in the major rivers at the northern and southern margins of the Kalahari, has exerted a considerable influence upon human activities and upon the development of the region in the colonial and post-colonial eras. Proposals devised in South Africa during the early twentieth century, to irrigate the Kalahari by diverting south the waters of the Zambezi and Okavango Rivers (Schwatz 1919, 1920), were never pursued but indicate the way in which the lack of surface water has been perceived as a hindrance to economic and agricultural development. Surface water does accumulate seasonally in pan depressions. These include both large features, such as the Makgadikgadi system and the km²-size depressions that are common in the south-western Kalahari, and small depressions of tens of square metres that can form in any shallow landscape low where clay particles may help seal the sand. These accumulations may last days, weeks, or even months into the dry season, and have been important foci for wildlife and their hunters and transhumance pastoralists throughout the region, with the presence of water contributing to the seasonal movements of animals and people alike.

Some of the dry valleys, or *mokgacha,* of the Kalahari may also flow following years of good rain. In the south-west, the Aoub may experience shallow flow every 3–5 years, while the Kuruman River flows for several tens of kilometres downstream of its spring source in most years.

Surface water deficiencies have inevitably given rise to quests to utilize groundwater resources instead. Boreholes have been sunk in the Kalahari at an increasing rate since the 1950s. Initially these were for drought relief purposes, but schemes to develop the Kalahari for livestock production have inevitably been based upon using groundwater to allow year-round access to grazing opportunities. The ever-growing rates of groundwater use raise issues of sustainability and recharge, though the lowering of Kalahari groundwater tables has been noted since the nineteenth century (Livingstone 1858; Thomas and Shaw 1991). With high rates of rainwater infiltration, the opportunities for recharge may be thought to be good. But with low rainfall levels and a thick sand cover, direct recharge is not likely (Foster *et al.* 1982). Rainfall in wetter locations on the rim of the Kalahari basin may offer the most likely source of recharge. DeVries (1984) and others have, however, noted, using isotope studies and analyses of the hydrological gradient, that currently ancient waters are being used and that significant aquifer replenishment has not taken place since 12,000 years ago. It would appear therefore that groundwater is being used unsustainably, with the likelihood of increasingly saline water extractions and ultimately of significant future borehole failures.

Soils, Plants, and Animals

The low content of basic nutrients (nitrate, potassium, and phospate) and the relative infertility of the sandy arenosols of the Kalahari have been widely commented upon (e.g. Skarpe and Bergstrom 1986; Buckley *et al.* 1987). Yet, even outside the atypical Okavango Delta with its permanent water availability, biomass levels are relatively high, and at least equal to other savanna areas in Africa and other semi-arid tropical and subtropical locations. This in part is now recognized to be a function of the relatively closed nature of Kalahari ecosystems, with plants, plant litter, and soil closely coupled, such that litter is rapidly broken down and the nutrients within made available for plant uptake.

There have been few systematic surveys of Kalahari-wide vegetation communities. Weare and Yalala's (1971) provisional (it has to date not been subsumed into a final version!) vegetation map of Botswana (Fig. 2.2) divides the Kalahari into three categories of shrub savanna in the drier south-western areas, eight categories of tree savanna in central, northern, and eastern areas, and four categories of grass savanna, which are largely within or adjacent to

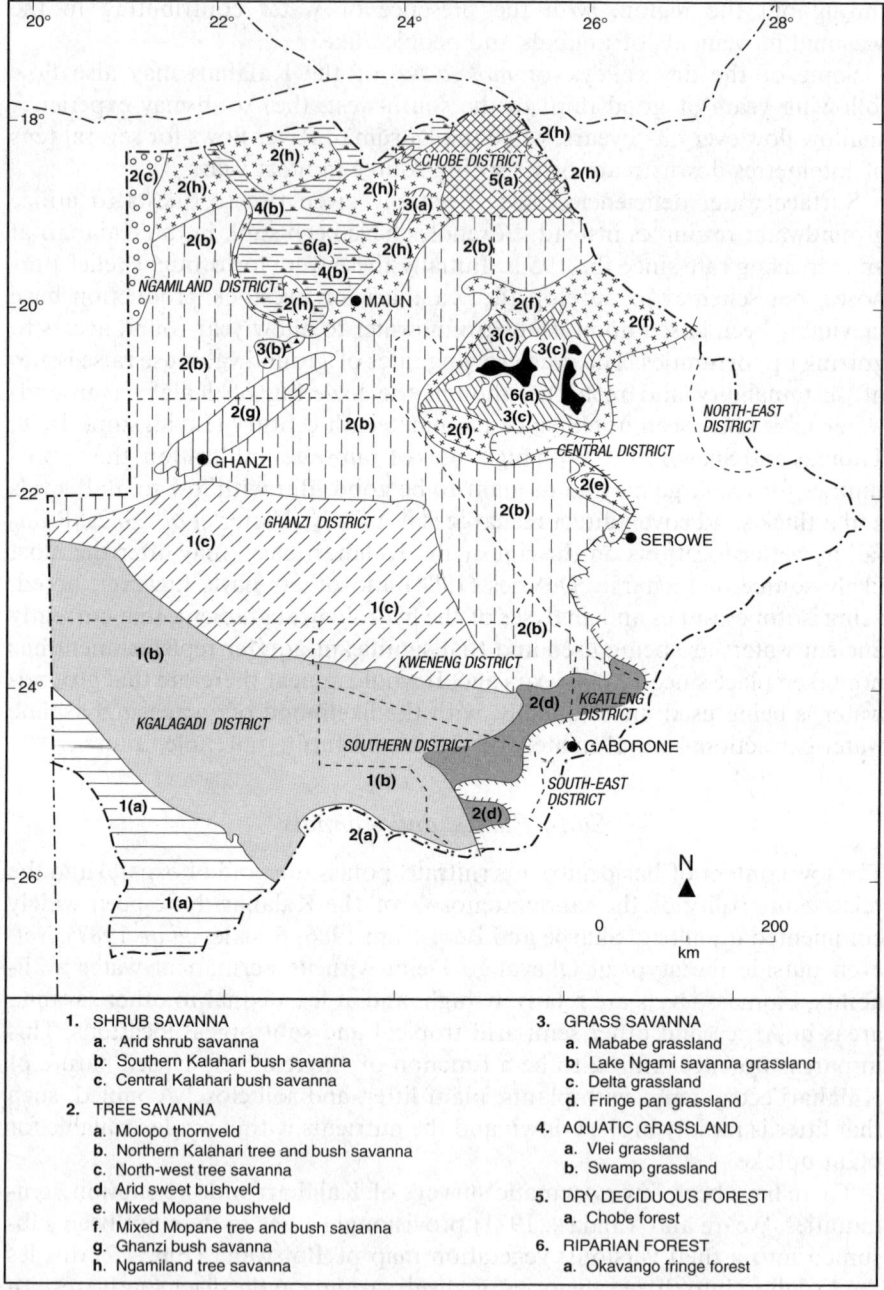

Fig. 2.2. Vegetation communities in the Kalahari, as defined by Weare and Yalala (1971).

major pan areas. The Okavango Delta supports aquatic grasslands and riparian forests. The terms shrub savanna and tree savanna are somewhat misleading, since in many situations grasses represent the dominant type of cover. Shrub savanna in the most arid areas, for example, generally comprises perennial grass species, especially in interdune areas, with occasional stunted shrubs. Fully grown trees, including tall camel thorns (*acacia erialoba*) and witgat (*boscia albitrunca*) trees, occur in dry watercourses and on dune flanks; both situations where stored water in sands can be tapped by extensive root systems. It is notable, however, that in the Molopo and Nossop *mokacha* of the south-west Kalahari, many of the largest channel floor trees are now dead, which may be indicative of declining water-table levels. Throughout the shrub savanna areas ground cover may be less than 35 per cent, since most grass species present are clumped rather than extensive. The tree and bush savannas found in the better watered central, northern, and eastern areas are again grass dominated, with low shrubs of *Acacia, Grewi*, and *Croton* species widespread and belts of more fully developed trees, again in situations where subsurface water opportunities are better. The one exception to this occurs in the areas of mopane woodland that are found north of 22°S. These are usually grassless, even when trees are not fully developed.

Within this simplified regional framework, local and sub-regional spatial variations in vegetation communities occur in response to four factors. First there are natural factors, including the depth of the water table, the presence of underlying calcareous or saline deposits, and so-called islands of fertility where soil nutrient levels are higher than average (Dougill *et al.* 1998). Second, around settlements, fuelwood collection may reduce the tree and shrub levels significantly, for distances of several kilometres. Third, fire, caused by lightning strikes and by humans, results in the removal of plant litter and can lead to the regeneration of grasses and suppression of new shrub growth.

Fourth, probably most significant and certainly the most commented upon, is herbivory (Skarpe 1991; Perkins and Thomas 1993*a*; Palmer and van Rooyen 1998). This may in particular lead to an increase in bare areas, especially close to water sources, and, further afield, to a change in the competitive relationship between the shrub/tree and grass layers. This contributes to what has been termed 'bush encroachment' by, for example, aggressive species such as *Acacia mellifera*. Such changes have frequently been regarded as effectively permanent and therefore as a component of desertification (e.g. Ringrose *et al.* 1990). Even though bush encroachment invariably represents an increase in biomass over previous mixed grass–shrub states, the loss of grass cover represents a decrease in the grazing resource, and therefore environmental degradation. Empirical and theoretical developments in the understanding of the operation of savanna ecosystems subjected to livestock practices have however led to a reappraisal of such views. It is recognized, for example, that livestock-induced changes have to be set against the natural

dynamics of dryland ecosystems (Ellis *et al.* 1993), which are driven primarily by rainfall variability. The identification of general changes in ecosystem
state, from a presumed norm or climax, is a redundant form of investigation
if regard is not given to the role of natural dynamics in contributing to
change. Where negative changes are created by livestock systems, their spatial
extent may be restricted by the radial pattern of range usage in borehole-
centred ranches. This leads to differential herbivore-use intensities (Perkins
and Thomas 1993*b*), with their long-term impact mitigated by complex linkages between plant and soil nutrients and soil water movements (Dougill
et al. 1998, 1999).

Since much of the Kalahari is unsuitable for cultivation, wild plants can
also provide an important direct component of the livelihoods of Kalahari
residents. Roots, bulbs, tubers, and berries are widely used on a seasonal basis
by rural people for food and medicinal purposes. Bush encroachment can
contribute to a reduction in biodiversity and a limitation of this supplementary dietary source.

Despite its reputation for wildlife, the dominant Kalahari mammal is the
cow. Although cattle have been grazed in the Kalahari for at least a thousand
years, and probably for longer when environmental conditions have been
favourable (Denbow 1984), the lack of permanent water limited their temporal occupancy. The spread of boreholes changed this, though under traditional livestock-rearing systems borehole-based cattleposts in the Kalahari
were still largely used only seasonally, when grassing resources permitted, to
relieve pressure in the eastern hardveld. Under such a system, competition
between cattle and wildlife, especially the migrating antelope species (hartebeest and wildebeest) and zebra, would have been limited. This situation
changed progressively during the 1970s and 1980s, however. First, with the
establishment of veterinary control fences, set up in response to demands to
control livestock movement in the face of foot-and-mouth disease outbreaks,
and second, with the progressively permanent occupancy of Kalahari borehole sites by cattle herders, particularly in association with the Tribal Grazing
Land Policy (TGLP) of 1975 (Tsimako 1991). The progressive fencing of
TGLP ranches during the 1990s (Thomas *et al.* 2000) has further contributed
to a decline in wildlife numbers, especially in central and northern areas. In
the south-western Kalahari of Botswana aridity and the poor quality of borehole water resources has limited the establishment of fenced TGLP ranches,
such that competition between livestock and wildlife, notably springbok and
gemsbok in this area, has been less marked. In the Northern Cape of South
Africa, however, fenced borehole-based farm units have existed for over
seventy years, leading to marked wildlife–cattle competition for much of the
twentieth century. Throughout areas that are used for livestock, the most
common antelope are the small solitary species such as duiker, steenbok, and
dik-dik.

The demise of wildlife numbers has impacted upon non-pastoralist resident groups, particularly the Khoisan, reducing opportunities for hunting and further compromising traditional livelihoods. The growth towards wildlife farming in some areas (see below), although decreasing reliance upon livestock, does not increase opportunities for these livelihoods since these areas, as with the national parks in the Kalahari, do not provide opportunities for subsistence hunting (or gathering) activities.

Kalahari Challenges

In this section some of the important issues that link the Kalahari environment and its use as a resource with the people who live in and attempt to extract a livelihood from the region, are explored. These issues are developed more substantially in other chapters in the book, but by considering them here in an environmental context, linkages between issues can be considered.

Managed or Unmanaged?

As the globalization of the countries that include the Kalahari increased during the later twentieth century, along with the development of their natural resources, it was inevitable that the Kalahari would progressively be viewed as a region to use rather than one to avoid. The notion of the Kalahari as an empty wilderness (or Eden) to avoid (or to leave alone), as propounded by van der Post (e.g. 1958) and others, was always false. Certainly through the last century the Kalahari has to one degree or another been managed. Different Khoisan hunter-gatherer clans had different territories (Hitchcock 1978), their extent leaving large areas unused at any time as a reflection of the extensiveness of this land-use activity. Transhumance pastoralism, with livestock entering peripheral parts of the Kalahari only when resources (water and grazing) permitted, was equally extensive in usage. The cattlepost (*meraka*) system employed in the Kalahari margins by Tswana groups from the mid-nineteenth century onwards had recognized rights of access to different areas by different groups, under the control of tribal authorities.

The notion of the Kalahari as empty and unused has increasingly been exploited by groups with more intensive aspirations for the region. The Northern Cape, including the Nossop, Molopo, and Kuruman valleys, and the Ghanzi Ridge of Botswana, were both mapped into farm units by European colonial authorities and assigned to white farmers in the late nineteenth and early twentieth centuries. It is no coincidence that it was these lands that were first divided; the *mokacha* being viewed as watercourses and the Ghanzi ridge being a zone of run-off and springs for geological reasons. Allocation ignored any ephemeral usage of these areas by hunter-gatherer groups, such that by the end of the nineteenth to the early twentieth century cattle were grazed, albeit extensively and in some areas only periodically, throughout the Kalahari by both Tswana and European groups (Denbow and Wilmsen 1986; Thomas and Shaw 1991). The one exceptional area to this was in the north, where the tsetse fly denied successful access by livestock.

From the 1950s onwards, coincident with the development of more efficient means of sinking and then pumping boreholes, an eye was being turned to the more intensive usage of the Kalahari for grazing (Debenham 1948, 1952). The formal development of the interior Kalahari for livestock production did not take place however until just before the end of the colonial period (Mazonde 1994) and especially in the post-colonial period, notably from 1975 onwards in association with the Tribal Grazing Land Policy. While the policy created opportunities for the controlled establishment of leasehold ranches in a number of blocks within the Kalahari, it has also been seen as creating opportunities for resource conflicts, environmental degradation, and the displacement of traditional livelihoods (White 1993). The conflicts relate to wildlife versus livestock, subsistence versus commercialism, rural resident versus distant urban lessee, and sustainability versus short-term gains. Although a lessee gains de facto control over water and therefore resources in a ranch, thereby denying access to others, many negative elements of the TGLP are not associated simply with the core of the policy. Rather, they relate to the failure to adhere to the policy, and to other temporally coincident changes. From an environmental perspective these include: the completion of a veterinary fence network, commenced in the 1950s, across the Kalahari, which inhibits wildlife movements across the region, contributing to the high mortalities in migrating herds in the 1980s (Williamson and Williamson 1985); the progressive fencing of ranch boundaries, particularly since 1991, constraining livestock to small areas, in deference to the operation of traditional cattleposts (Perkins 1996); and the sinking of extra boreholes at a spacing less than the permitted one every 8 km.

The regulation of the Kalahari through management is not therefore new, but in recent decades it has seen increasing centralization and intensification. Conflicts between livestock and wildlife, and environmental degradation due to livestock production, were both noted in the Kalahari as long ago as the late nineteenth century (see Thomas and Shaw 1991), as was the displacement of traditional livelihoods by new arrivals.

Two general management developments have impacted particularly significantly on the Kalahari environment (Yeager 1989): the protection of wildlife (the first Kalahari national park was established in 1930) and the commercialization of livestock production. In Western eyes one is commonly seen as creating a (people-less) Eden, the other an environmental Armageddon. Sandwiched between the two has been the demise of the land-use system that is seen as most sustainable. This is now to some extent catered for, in Botswana, in Wildlife Management Areas (WMAs) (see Ch. 7). What all these management changes have meant is an increasing fragmentation of the Kalahari into units that are sometimes too small to cope, without further external intervention, with the natural environmental variability that affects the Kalahari region.

Crops, Livestock, and Wildlife: Staying Still or Moving On

Although commercial crop production is carried out intensively in a number of areas around the Kalahari margins (for example, grain and peanut production in south-east Botswana and North West Province, South Africa, and viticulture on the banks of the Orange River around Upington), the climate and soils of the Kalahari make the region generally unsuitable for large-scale agriculture. This has of course been recognized by African and European groups alike, leading to a focus upon cattle production amongst those seeking to make commercial gains from the area, with cultivation usually restricted to small fields or plots used for growing staple crops for local use. Cattle require water on a daily basis, so that efforts to enhance Kalahari cattle production, including under the TGLP, have inevitably led to a heavy use of groundwater, and the environment surrounding water-points. Even with the use of borehole water, susceptibility to drought is not overridden, since drought impinges upon the growth of vegetation that is necessary as food. This occurs even if recommended stocking levels have been adhered to, since these tend to be based upon average conditions, which are rarely met in the climatically dynamic Kalahari. Drought-induced livestock die-offs are not therefore prevented by boreholes since cattle starve to death during long droughts, unless expensive fodder supplements are brought on to the ranches, or cattle are moved off drought-affected areas. The latter strategy was noted to be in use, for example, in the Ncojane ranch area of the western Kalahari during the 1994–5 drought (Thomas *et al.* 2000), when livestock was moved from ranches to cattleposts up to 50 km away This strategy increases food availability, and indicates the use of traditional practices, in deference to those of the TGLP, at times of environmental stress.

Access to alternative lands during stress periods is not available to all livestock producers in the Kalahari. In some areas, the consolidation of holdings, to increase the available range above the amount set out in the original demarcation of units, has taken place. This is particularly noticeable amongst the European farms in the Kalahari of northern South Africa, where in many instances adjacent units in the Molopo valley have been combined into one farm. An alternative strategy, which is seen as offering a number of benefits, is to replace livestock with managed wildlife (e.g. Cooke 1985). This is increasingly being employed in parts of the Kalahari, where farmed game demands less water per head than cattle, grazes less selectively, and is more drought resistant. In some areas game farming is supplemented by the sale of hunting licences, and is employed in integrated management in WMAs in Botswana (though these are not without their own problems) and in community projects in the Northern Cape of South Africa. Replacing cattle with wildlife is not necessarily the panacea for livestock-related problems in the Kalahari, for example, because suitable markets have to be found for wildlife meat to make it commercially viable and an acceptable replacement for cattle. The greater drought tolerance and more catholic dietary habits of species such as eland compared to cattle have led many authors (e.g. Pollock 1969; Cooke 1985) to view game ranching as more environmentally sensitive, probably more productive, and better at meeting the commercial and environmental demands of modern states that impinge on the Kalahari.

Kalahari Degradation . . . or Natural Variability

The nature of the savanna ecosystems of the Kalahari means that temporal and spatial variability are normal characteristics of vegetation at any snapshot of time. Two principal factors contribute to this variability: animals and climate. The interactions between grazing and browsing animals and vegetation are complex in savanna systems (Walker 1987). On the one hand, intense grazing can lead to excessive removal of the most edible species, which are usually perennial grasses. This reduces ground cover, but ultimately opens the way for less palatable annual grasses to gain a foothold, and, in many situations in the Kalahari, for shrub and bush species to become dominant. Alternatively, just enough grazing pressure stimulates growth by the palatable species, and increases their biomass (Crawley 1983) and nutrient content (Scholes 1990). Livestock systems, especially sedentary ones unless they are managed intensively, have limited opportunities for the natural regulation of grazing pressures compared to unbounded ones where animals can and do move through the landscape grazing selectively, thereby diminishing the likelihood of excessive pressures on vegetation.

Grazing and climatic variability affect the amount of ground cover and the ratio between different plant elements, for example, shrubs, perennial grasses, and annuals. Climatic variability, both seasonally and interannually, affects which species are senesced or actively growing. Plants are opportunistic, with small-scale spatial differences reflecting factors including local soil nutrient differences, soil depth, and water availability. At larger scales, the spottiness of rainfall events contributes to biomass differences and provides a link to the temporal variability of biomass and plant growth. The elements of variability mean that vegetation does not progress through stages dominated by different plant classes or species towards an ultimate climax, but rather that dynamism, both 'up' in response to growth favouring conditions and 'down' in response to negative factors such as moisture shortages, fire, and heavy grazing, are the norm. This 'state and transition' (Walker et al. 1981) or 'disequilibrium' (Behnke et al. 1993) behaviour means that reductions in diversity and biomass are not necessarily an indication of the fragility of such systems, but of the strategies and adaptations employed to cope with less favourable circumstances and to set store for the onset of more favourable conditions in the future.

Distinguishing natural variability from the enhanced levels of change that may result from excessive pressures related to human actions (and from which recovery may not ensue if favourable conditions are restored) can create problems in the assessment of degradation. The duality of increased vegetationless areas (sacrifice zones) and areas experiencing bush encroachment is widely cited as the principal characteristic of Kalahari environmental degradation. In many parts of the Kalahari the distribution of sacrifice and bush-encroached zones is related to the radial pattern of livestock pressures around boreholes (e.g. Perkins 1996). This means that the extent of these zones is unlikely to increase over time unless more systems become degraded, or new boreholes are instigated on existing grazing units (as long ago as 1971, a spacing of 8 km between boreholes in the Kalahari, each supporting no more than 400 head of livestock, was recommended by Jarman and Butler 1971), thereby increasing the number of points from which pressure originates.

Bush-encroached zones can in theory decrease in extent, through the effects of natural fire or specific management practices that restore the ecological balance in favour of grasses, since associated changes (nutrient and water) in the soil resource of encroachment-affected areas are generally absent (Dougill *et al.* 1999). Thus environmental restoration, at least in theory, is attainable. Few examples presently exist of specific attempts to decrease the extent of the bush layer, though in the Meir area of the Northern Cape, a community-based bush clearance and grass replanting scheme has been assessed, on lands where cattle production is being replaced by integrated livelihoods involving wildlife (van Rooyen 1998). Drought events may not appreciably reduce bush densities. Encroached areas may be relatively drought resistant (compared to grass-rich areas) due to the networks of deeper-rooting shrubs, a view supported by interannual analyses of vegetation cover, using remotely sensed data and field studies, by Ringrose *et al.* (1990). Finally, it can be noted that human interference with wildlife populations and their areas of concentration means that they too can contribute to the enhanced spatial variability of Kalahari vegetation. This has been monitored, for example, within the Kalahari–Gemsbok National Park (now the Kalahari Transfrontier Park), and in particular relates to the provision of dry-season water-points, especially within the dry watercourses of the Nossop and Auob valleys (van Rooyan *et al.* 1984).

Global Warming and Sustainable Resource Use

High levels of interannual rainfall variability and the relatively frequent occurrence of drought are normal in the Kalahari. Wildlife and traditional livelihoods are both adapted to these characteristics. Wildlife and traditional livelihoods have also declined in the Kalahari in the last two decades, with their common replacements, cattle and sedentary livelihoods, having a greater vulnerability to their impacts. This susceptibility has been reflected both in the increased reliance of vulnerable social groups on drought relief activities and food distributions (Sporton *et al.* 1999) and in the relative failure of the TGLP to establish large-scale cattle production in the Kalahari successfully (White 1993). Where drought is coped with relatively well on TGLP ranches it has been found to be a consequence of the full package of TGLP conditions not having been applied and the reversion of pastoralists to traditional Kalahari cattlepost strategies (Thomas and Sporton 1997).

The annual mean and interannual variability of rainfall in the Kalahari are not consistent, and vary at the decadal and greater scales. Tyson's (1985) quasi-18-year drought cycle for the southern African rainfall zone is well known, but longer trends are evident too. In an Africa-wide comparison of rainfall in the periods 1931–60 and 1961–90, Hulme (1992) identified a decline of about 5 per cent in annual rainfall amounts in the tropical margins of southern Africa, which includes the heart of the Kalahari. Recently attention has turned to the impact that probable anthropogenically enhanced global warming may have on the Kalahari in the twenty-first century. Limitations in General Circulation Models and the relative paucity of analogue data to input for tropical and subtropical regions probably leaves predictions for tropical and subtropical areas as even more imprecise than those for temperate latitudes. Joubert (1995) reviewed available GCMs for southern Africa, with Joubert *et al.* (1996) then using that deemed the most reliable to model the likely impacts of double CO_2 climates on southern African droughts. They suggested that mean annual rainfall totals are unlikely to change significantly in the Kalahari, but that the number of small rainfall events (less than 3 mm a day) would decline, while the number of days with over 113 mm of rainfall might increase by over 40 per cent. Linked to this is an increase in the number of rain days in the middle of the summer wet season, and a decline in the frequency of periods with five or more consecutive rainless days. Overall, a decline in the probability of drought years was predicted.

Joubert *et al.*'s (1996) modelling predicts that the Kalahari will not be adversely affected by global warming, and may even benefit from more reliable rainfall characteristics. Other models suggest a different outcome. Joubert *et al.* (1996) did not consider the effect that temperature increases might have on factors such as soil moisture. In another study, using a different GCM, Hulme (1996) predicts a 2 °C increase in Kalahari mean air temperatures by 2050, increasing evapotranspiration 4–12 per cent. To model rainfall changes, Hulme (1996) used a core GCM and two others that respectively predict greater wetting and greater drying than the core model. The core model predicts little change to present rainfall in the Kalahari, as is the case with the dry model. The wet model suggests an increase in annual precipitation of over 7.5 per cent. The incidence of El Niño events leading to drought was not predicted to change.

Notwithstanding the limitations of GCMs, overall rainfall in the Kalahari is widely believed to change little from present conditions; however, the levels of available moisture are likely to fall due to the impact of temperature increases, with soil moisture declining below its already limited availability. Applying this to Prentice *et al.*'s (1992) biome model, the shrub savanna of the south-west Kalahari is predicted to increase in extent by 30 per cent, at the expense of the higher biomass savannas found to the north.

Kalahari Futures

The future state of the Kalahari environment and its utility for different livelihoods is uncertain due to both environmental and socio-economic factors. Ecological changes during the last decades of the twentieth century have resulted from the increased sedentarization of land-use practices and from the fragmentation of the land. These give rise to fewer opportunities for the environment to recover from the effects of natural drought events, and a greater dependence of human and livestock groups on relatively small, clearly defined, areas of land. Current predictions of the impact of anthropogenically enhanced global warming on the Kalahari are relatively benign compared to those for other dryland regions of the world (e.g. Williams and Balling 1995), though temperature increases alone will reduce soil moisture levels. It is tempting to suggest, given the levels of social and political intervention in the Kalahari environment since the 1970s, that by 2050 the impacts of policies affecting land use in Kalahari countries will have been the dominant determinant of the state of the Kalahari in the twenty-first century, affecting the rate and extent of land degradation (or its mitigation) and the distribution and size of the groups who have access to the environmental resources of the Kalahari.

References

ARNTZEN, J. W., and VEENENDAAL, E. M. (1986), *A Profile of Environment and Development in Botswana* (Gaborone: NIR).

BEHNKE, R. H., SCOONES, I., and KERVEN, C. (eds.) (1993), *Range Ecology at Disequilibrium* (London: ODI).

BHALOTRA, Y. P. R. (1985), *Rainfall Maps of Botswana* (Gaborone, Botswana: Department of Meteorological Services).

BUCKLEY, R., WASSON, R., and GUBB, A. (1987), 'Phosphorus and potassium status of arid dunefield soils in central Australia and southern Africa, and biogeographic implications', *Journal of Arid Environments*, 13: 211–16.

COOKE, H. J. (1985), 'The Kalahari today: A case of conflict over resource use', *Geographical Journal*, 151: 75–85.

CRAWLEY, M. J. (ed.) (1983), *The Dynamics of Animal–Plant Interactions* (London: Blackwell), 437, s.v. Herbivory.

DEBENHAM, F. (1948), 'Report on the resources of Bechuanaland, Nyasaland, Northern Rhodesia, Tanganyika, Uganda and Kenya', *Colonial Research Publication*, 2: 31–9.

——(1952), 'The Kalahari today', *Geographical Journal*, 118: 12–23.

DENBOW, J. R. (1984), 'Prehistoric herders and foragers of the Kalahari: The evidence for 1500 years of interaction', in C. Schrire (ed.), *Past and Present in Hunter-Gatherer Studies* (Orlando: Academic Press), 175–93.

DENBOW, J. R., and WILMSEN, E. N. (1986), 'Advent and course of pastoralism in the Kalahari', *Science*, 234: 1509–15.

DEVRIES, J. J. (1984), 'Holocene depletion and active recharge of the Kalahari groundwaters—a review and an indicative model', *Journal of Hydrology*, 70: 221–32.

DOUGILL, A. J., HEATHWAITE, A. L., and THOMAS, D. S. G. (1998), 'Soil water movement and nutrient cycling in semi-arid rangeland: Vegetation change and system resilience', *Hydrological Processes*, 12: 443–59.

DOUGILL, A. J., THOMAS, D. S. G., and HEATHWAITE, A. L. (1999), 'Environmental change in the Kalahari: Integrated land degradation studies for nonequilibrium dryland environments', *Annals, Association of American Geographers*, 89: 420–42.

ELLIS, J. E., COUGHENOUR, M. B., and SWIFT, D. M. (1993), 'Climatic variability, ecosystem stability and the implications for range and livestock development', in R. H. Behnke, I. Scoones, and C. Kerven (eds.), *Range Ecology at Disequilibrium* (London: ODI), 31–41.

FOSTER, J. B., BATH, A., FARR, J., and LEWIS, W. (1982), 'The likelihood of active groundwater recharge in the Botswana Kalahari', *Journal of Hydrology*, 55: 113–36.

HITCHCOCK, R. K. (1978), *Kalahari Cattle Posts: A Regional Study of Hunter-Gatherers, Pastoralists and Agriculturalists in the Western Sandveld Region, Central District, Botswana* (Gaborone: Government Printer).

HULME, M. (1992), 'Rainfall changes in Africa: 1931–60 to 1961–90', *International Journal of Climatology*, 12: 685–99.

——(ed.) (1996), *Climate Change in Southern Africa: An Exploration of some Potential Impacts and Implications in the SADC Region* (Norwich: University of East Anglia, CRU, and Gland: WWW International).

JARMAN, T. R. W., and BUTLER, K. E. (1971), 'Livestock management and production in the Kalahari', *Botswana Notes and Records, Special Edition*, 1: 132–9.

JOUBERT, A. M. (1995), 'Simulations of southern African climate using early-generation general circulation models', *South African Journal of Science*, 91: 332.

JOUBERT, A. M., MASON, J., and GALPIN, J. S. (1996), 'Droughts over southern Africa in a double-CO_2 climate', *International Journal of Climatology*, 16: 1149–56.

LIVINGSTONE, D. (1858), *Missionary Travels and Researches in South Africa* (London: Deighton Bell).

MAZONDE, I. N. (1994), *Ranching and Enterprise in Eastern Botswana* (Edinburgh: Edinburgh University Press).

PALMER, A. R., and VAN ROOYEN, A. F. (1998), 'Detecting vegetation change in the southern Kalahari using Landsat TM data', *Journal of Arid Environments*, 39: 143–53.

PERKINS, J. S. (1996), 'Botswana fencing out the equity issue. Cattleposts and cattle ranching in the Kalahari Desert', *Journal of Arid Environments*, 33: 503–18.

PERKINS, J. S., and THOMAS, D. S. G. (1993*a*), 'Spreading deserts or spatially confined environmental impacts? Land degradation and cattle ranching in the Kalahari Desert of Botswana', *Land Degradation and Rehabilitation*, 4: 179–94.

——(1993*b*), 'Environmental responses and sensitivity to permanent cattle ranching in semi-arid western central Botswana', in D. S. G. Thomas and R. J. Allison (eds.), *Landscape Sensitivity* (Chichester: Wiley).

PIKE, J. G. (1971), 'The development of the water resources of the Okavango Delta', *Botswana Notes and Record, Special Edition*, 1: 35–40.

POLLOCK, N. C. (1969), 'Some observations on game ranching in southern Africa', *Environmental Conservation*, 2: 18–25.

PRENTICE, J. C., CRAMER, W., HARRISON, S. P., LEEMANS, R., MONSEERUD, R., and SOLOMON, A. M. (1992), 'A global biome model based on plant physiology and dominance, soil properties and climate', *Journal of Biogeography*, 19: 117–34.

RINGROSE, S., MATHESON, W., TEMPEST, F., and BOYLE, T. (1990), 'The development and causes of range degradation features in south-east Botswana using multi-temporal Landsat MSS imagery', *Photogrammatic Engineering and Remote Sensing*, 56: 1252–62.

SCHOLES, R. J. (1990), 'The influence of soil fertility on the ecology of dry savannas', *Journal of Biogeography*, 17: 415–19.

SCHWATZ, E. H. L. (1919), 'The progressive desiccation of Africa: The cause and the remedy', *South African Journal of Science*, 15: 139–90.

——(1920), *The Kalahari, or Thirstland Redemption* (Cape Town: Maskew Miller).

SKARPE, C. (1991), 'Impact of grazing in savanna ecosystems', *Ambio*, 20: 351–6.

SKARPE, C., and BERGSTROM, R. (1986), 'Nutrient content and digestability of forage plants in relation to plant phenology and rainfall in the Kalahari, Botswana', *Journal of Arid Environments*, 11: 147–64.

SPORTON, D., THOMAS, D. S. G., and MORRISON, J. (1999), 'Outcomes of social and environmental change in the Kalahari of Botswana: the role of migration', *Journal of Southern African Studies*, 25: 441–59.

THOMAS, D. S. G. (1986), 'Dune pattern statistics applied to the Kalahari Dune Desert, southern Africa', *Zeitschrift für Geomorphologie*, 30: 231–42.

——(1988), 'Environmental management and environmental change: The impact of animal exploitation on marginal lands in central southern Africa', in J. C. Stone (ed.), *The Exploitation of Animals in Africa* (Aberdeen: Aberdeen University), 5–22.

THOMAS, D. S. G., and SHAW, P. A. (1991), *The Kalahari Environment* (Cambridge: Cambridge University Press).

THOMAS, D. S. G., and SPORTON, D. (1997), 'Understanding the dynamics of social and environmental variability. The impacts of structural land use change on the environment and peoples of the Kalahari, Botswana', *Applied Geography*, 17/1: 11–27.

THOMAS, D. S. G., SPORTON, D., and PERKINS, J. S. (2000), 'The environmental impact of livestock ranches in the Kalahari, Botswana: Natural resource use, ecological change and human response in a dynamic dryland system', *Land Degradation and Development*, 11: 327–41.

TSIMAKO, N. (1991), *Tribal Grazing Land Policy (TGLP) Ranches. Performance to Date* (Gaborone: Ministry of Agriculture).

TYSON, P. D. (1979), 'Southern African rainfall, past present and future', in M. T. Hinchley (ed.), *Proceedings of the Symposium on Drought in Botswana* (Gaborone: Botswana Society), 45–52.

——(1986), *Climate Change and Variability in Southern Africa* (Oxford: Oxford University Press).

VAN DER POST, L. (1958), *The Lost World of the Kalahari* (London: Hogarth).

VAN ROOYEN, A. F. (1998), 'Combating desertification in the southern Kalahari: Connecting science with community actions in South Africa', *Journal of Arid Environments*, 39: 285–98.

VAN ROOYEN, N., VAN RENSBURG, D. J., THERON, G. K., and BOTHMA, J. DU P. (1984), 'A preliminary report on the dynamics of the vegetation of the Kalahari Gemsbok National Park', *Koedoe Supplement*, 83–102.

WALKER, B. H. (ed.) (1987), *Determinants of Tropical Savanna* (Oxford: IRL Press).

WALKER, B. H., LUDWIG, D., HOLLING, C. S., and PETERMAN, R. S. (1981), 'Stability of semi-arid savanna grazing systems', *Journal of Ecology*, 69: 473–98.

WEARE, P. R., and YALALA, A. (1971), 'Provisional vegetation map', *Botswana Notes and Records*, 3: 131–52.

WHITE, R. (1993), *Livestock Development and Pastoral Production on Communal Rangeland in Botswana* (Gaborone: Botswana Society).

WILLIAMS, M. A. J., and BALLING, R. C. (1995), *Interactions of Desertification and Climate* (London: Edward Arnold).

WILLIAMSON, D., and WILLIAMSON, J. (1985), 'Botswana's fences and the depletion of Kalahari wildlife', *Parks*, 10: 5–7.

YEAGER, R. (1989), 'Demographic pluralism and ecological crisis in Botswana', *Journal of Developing Areas*, 23: 385–404.

3

Politics, Policy, and Livelihoods

Deborah Sporton and Chasca Twyman

Introduction

Contemporary development discourses recognize the need to move away from top-down towards bottom-up approaches to poverty alleviation, where the poor are viewed as stakeholders rather than victims in the policy-making process (World Bank 2000). In the implementation of policy, community consultation and grass-roots involvement are encouraged with the aim of empowering the poor. Within the Kalahari of southern Africa which is today incorporated within the policy domains of three nations, Botswana, Namibia, and South Africa, a key contemporary policy challenge is the need to respond to the legacy of colonial and subsequent apartheid land and agricultural policies that have created an underclass of rural poor who are deprived of the capital assets necessary to sustain livelihoods. This chapter examines very different approaches to land and agricultural reform in Botswana, Namibia, and South Africa and the implications and challenges they represent for rural natural resource-based livelihoods and for local empowerment in the Kalahari region. The research findings presented here are focused on three cross-border areas spanning North West Province, South Africa and Southern District, Botswana; Ghanzi District, Botswana and Omaheke District, Namibia; Kgalagadi District, Botswana and Northern Cape Province, South Africa (Fig. 3.1). These study areas provide different environmental and policy contexts for the investigation of natural resource use and rural poverty.

Historical Frameworks

Contemporary policy interventions and enviro-social livelihood systems in the Kalahari represent the outcome of complex narratives of occupation, differentiation, territorial conflict, and appropriation, mapped onto a long history of environmental change and cyclical drought. Behind the rhetoric of

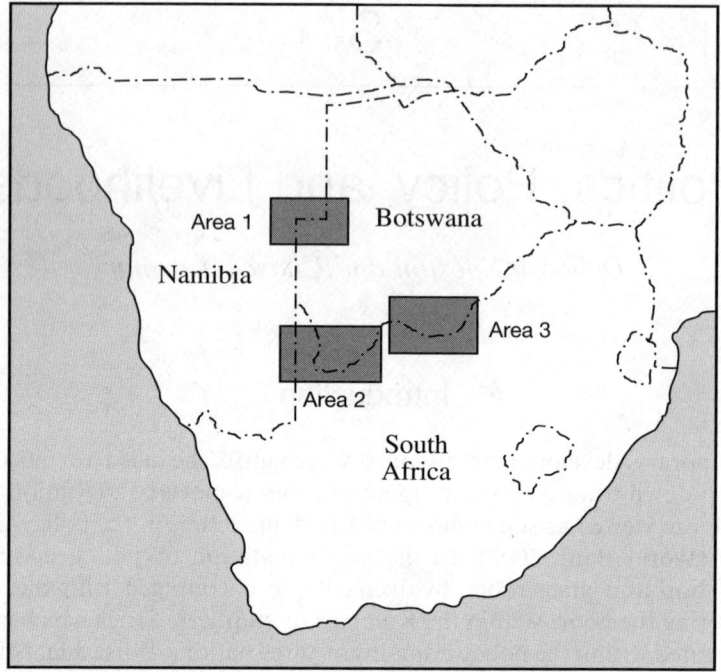

Fig. 3.1. The three Kalahari cross-border study areas.

policy lies the historical context in which enviro-social relations are embedded and discursively produced. Wilmsen (1989) argues that dialectical relationships are often ignored in favour of particularism with the focus on one particular group or one particular point in time. As a result, 'the history of the Kalahari as written, reads like a kaleidoscope of unconnected slide shows thrown up on segregated screens' (p. 7). While it is beyond the scope of this chapter to detail this history in depth, this section attempts to link these 'slides' encompassing pre-colonial, colonial, and post-colonial periods, to emphasize key moments in the historical production of contemporary natural resource-based livelihoods and policy in the Kalahari region.

Authenticity and Pre-Colonial Livelihood Systems

One such distortion in the understanding of contemporary livelihoods concerns the conventional wisdom that pre-colonial indigenous livelihood systems in the Kalahari have endured the passage of time. Until recently, the first peoples of the Kalahari, the San (often referred to as Basarwa), romanticized as 'bushmen' in the popular literature (van der Post 1958) and as pristine historical artefacts in the academic literature (see for example Lee and

DeVore 1968), were believed to have existed in relative isolation employing adaptive foraging livelihoods to cope with the harsh Kalahari environment. It is now recognized that as long as 2,000 years ago, with the introduction of cattle into the Kalahari, the San became active participants in a developing pastoral economy with Bantu-speaking peoples, primarily Tswana and Herero groups (Wilmsen 1989). During the eighteenth and nineteenth centuries, with the continued expansion of Tswana agro-pastoralism into the southern and eastern Kalahari, pre-existing indigenous social relationships were transformed as emerging Tswana states imposed clientage relationships on the San (Wilmsen 1989; Wylie 1994). The legacy of this pre-colonial system of servitude and hierarchical social relations endures today with Tswana political hegemony and the marginalization of the San in virtually all spheres (Good 1993; Motzafi-Haller 1998).

From Colonization to Independence, Legacies, Livelihoods, and Implications for Contemporary Policy

Pre-colonial structures were reinforced by colonial expansion into southern Africa during the second half of the nineteenth century when the Kalahari region and its nascent Tswana kingdoms were incorporated into the British Bechuanaland Protectorate created in 1885 and German South West Africa established in the 1890s, with the southern margins of the Kalahari annexed into the Union of South Africa in 1910 by the British. With territorial rule began the rather arbitrary imposition of controlled borders dividing the Kalahari and thereby disrupting tribal territories, villages, and kinship networks and the organization of local livelihoods (see, for example, Manson and Drummond 1991). The enforcement of colonial rule in these new territories, at least in its early stages, had different impacts on the peoples and environment of the Kalahari.

In the Bechuanaland Protectorate, the first fifty years of British indirect rule was administered through the Tswana chiefdoms that were able to consolidate their position at the expense of other groups, most notably the San, who were subjugated for their labour. Only in the 1930s with a change in administration (Parsons and Crowder 1988), did the emphasis of colonial policy change in favour of development, specifically water development in the Kalahari, following calls from Tswana élite who, as large cattle-owners, were keen to exploit this untapped grazing reserve (Schapera 1943). The extensive sinking of boreholes that ensued also transformed property rights in the Kalahari, traditional tribal management of boreholes was replaced with a system of syndicated ownership which favoured the political élite, the large cattle-owners, many of whose descendants retain control to the present (Hitchcock 1985; Peters 1994). The impact of cattle production on the Kalahari environment was regarded as negligible by scientists of the day further adding credence to this programme of expansion (Debenham 1948). At independence in 1966, Botswana inherited a burgeoning, yet largely unregulated, cattle industry expanding westwards deep into the Kalahari and pressuring finite grazing resources (Hitchcock 1985). Other sectors of the economy were severely undeveloped, the legacy of colonial neglect, making Botswana one of the poorest countries in the world (Yeager 1989). However in just thirty years, the country's economic fortunes have been transformed initially through the development of export-led cattle production and now, significantly, through the commercial exploitation of rich mineral reserves, particularly diamonds. Botswana has emerged as one of the dominant economic powers in sub-Saharan Africa and while per capita income levels remain high, they mask extreme inequalities in wealth between rich and poor (Good 1992; Mayende 1993). This reflects both historically rooted social and political divisions and 'modernizing' post-colonial legislative reforms. Charged by central government, a succession of policies has been implemented to increase the productivity and commercial viability of the Kalahari environment often to the detriment of local livelihoods.

In the Union of South Africa, by contrast, the southern Kalahari was developed at the outset by the British for commercial farming along the Molopo river, in Gordonia and the Northern Cape. Here native populations had existed in various sharecropping and tenancy arrangements with white settlers until the 1913 Natives' Land Act disbarred them from farming 92 per cent of the land and consigned Africans to native reserves constituting the remaining 8 per cent of the land area. This reserved area was later extended to 13 per cent as severe overcrowding, degradation, and poverty raised the prospect of acute subsistence crises. The act transformed the system of agricultural production on the margins of the Kalahari. It became one where Africans, deprived of their livelihoods, were either pushed into labour tenancy and then into a system of fettered waged labour or were forced to migrate to work elsewhere, either in the South African mines or in the cities (Bernstein 1996). From 1948, segregation became institutionalized through the apartheid regime which initially aimed to stabilize the growing influx of black Africans to urban manufacturing centres and to control the loss of farm labour from rural areas through the institution of Pass Laws governing conditions of residence. The 1950s saw the acceleration of apartheid legislation covering virtually all spheres of social and economic life. In 1959 the Bantu self-government Act resulted in the creation of ten ethnic homelands of which all Africans were to become citizens—between 1960 and 1983 it is estimated that 3.5 million Africans were forcibly removed to the homelands from white farms, black freehold farms (owned before 1913), and urban areas (Platzky and Walker 1985). Of significance to the political economy of the Kalahari region was the creation of Boputhatswana, a fragmented state on the southern margins of the Kalahari and the periphery of white South Africa which was granted nominal independence in 1977 (Jones 1999). In line with the homeland policy which aimed to encourage a 'national culture',

Boputhatswana, particularly post-independence, was founded on the revival of Tswana nationalism; indeed its name means 'the place where Tswana people reside'. Under the leadership of Louis Mangope, administrative structures were established, building on traditional customary authorities, with the aim of modernizing the former homeland and seeking, without success, legitimation from the international community (Lawrence and Manson 1994). With the collapse of the Bantustans in 1994, and the incorporation of Boputhatswana into North West Province, the effect of customary discourses and practices and the role of tribal authorities in land allocation have remained a potent force influencing the politics of land and rural livelihoods (Bernstein 1998).

Colonization in the Kalahari of German South West Africa was more coercive and allied to breaking local ties to the land resulting in the confiscation of Herero cattle and grazing lands and ultimately in the large-scale out-movement of Herero into the Bechuanaland Protectorate following the German–Herero war of 1904. In 1915, the territory of South West Africa was transferred to the Union of South Africa heralding the onset of large-scale white settlement of the western Kalahari by Afrikaans-speaking whites many of whom had been dispossessed during the Anglo-Boer war (1899–1902). In accordance with the 1913 Natives' Land Act, large areas of the Kalahari region were demarcated for white settlement only, smaller areas were designated as native reserves, notably the Amenuis and Epikuro (later known as Hereroland East) reserves which initially accommodated Tswana and Herero groups respectively. The discourse legitimizing this policy was that 'natives' should be provided with communally owned land where they could practise traditional subsistence activities thereby allowing white settlers to realize the commercial potential of the richest farming land. The San, dominated by the Ju'/hoansi group, who had hitherto coexisted in a servile relationship with Tswana and Herero pastoralists, now found themselves excluded in the allocation of reserves and deprived of the lands that had sustained their foraging livelihoods (Gordon 1992). While white settlers established heavily subsidized commercial livestock farms, communal farmers on the less viable, marginal reserve lands, where grazing and water were limited, were forced to sell their labour to support their livelihoods. At the time of Namibian independence in 1990, the political economy of the western Kalahari was dominated by white commercial farmers who were thriving at the expense of communal farmers unable to subsist, excluded from the wider economy, and, in the case of the San, consigned to the underclass. The Namibian government has been left facing policy conflicts between radical land reform and redistribution thereby undoing the wrongs of the past, and reconciliation, appeasing white commercial farmers whose production makes a significant contribution to the national economy (Tapscott 1993). Additional pressure to preserve the status quo has come from the World Bank who advised that land redistribution would adversely affect rural productivity and employment (World Bank 1991). Pankhurst (1995: 551) has argued that the failure to make sweeping land reforms owes much to 'the prevalence of a conservative colonial intellectual legacy'.

The colonial legacy has undoubtedly shaped contemporary policy debates and the social production of livelihood entitlements in the Kalahari of post-independence southern Africa. Historical pre-colonial relations of power founded on ethnicity were reinforced under colonial rule and, in the case of South Africa and Namibia, were further politicized through the policies of apartheid. At independence, in Botswana, Namibia, and South Africa, rural poverty was endemic with access to land and environmental resources restricted for certain groups. The environmentally marginal communal lands of the Kalahari, in which the majority of the rural poor were concentrated, were overgrazed, resource-poor, and degraded. The policy challenge has been to redress the injustices of the past yet, at the same time, balance demands for social equity and improved livelihoods for the rural poor with the need for economic growth and consequent implications for the preservation of the status quo. As a result, key post-colonial policy initiatives in the region have combined land reform with attempts to commercialize rural production. The implications of these national land policies for local livelihoods, which are the focus this chapter, can therefore only be understood by examining the historical context in which they are grounded.

Land Reform and Land Policies

Within southern Africa, traditional, strong, dualistic tenure structures have not changed significantly since independence. Commercial production dominates the indigenous or 'communal' sector which has, for the most part, been ignored or used as reserve land to support the commercial sector in times of need. Three cases are presented in this section, each of which highlights a different type of land reform policy in southern Africa: restitution in South Africa, resettlement in Namibia, and settlement in Botswana. For each, the focus is on how such policy change will affect livelihood options for rural people living within Kalahari environments. Both the constraints and opportunities offered by these policy changes will affect the short- and long-term viability of livelihoods, particularly those based on natural resources.

Land Restitution in South Africa

Land restitution in South Africa aims to restore land and provide other remedies to people dispossessed by racially discriminatory legislation and practice. Restitution is an integral part of South Africa's broader land reform programme and is closely linked to the redistribution of land and tenure reform programmes (Department of Land Affairs 1997). Land reform in South Africa is central to the country's aims of reconciliation, reconstruction, and development in the post-apartheid era. It is closely and explicitly linked with the enhancement of sustainable livelihoods with aims to reduce poverty, diversify sources of income, and allow people more control over their lives and their environment, and is further expected to reduce the risk of land degradation.

The government's policy and procedure for land claims (the basis of restitution) are based on the provision of the Constitution and the Restitution of Land Rights Act 22 of 1994. Claims arising from dispossession after 19 June 1913 (Natives' Land Act) are handled by the Land Claims Commission and Land Claims Court, while claims prior to 1913 are redirected to the Land Distribution Programme and are classified as historical claims. Land restitution can take the form of (a) restoration of the land from which claimants were dispossessed, (b) provision of alternative land, (c) alternative relief comprising a combination of the above, or (d) priority access to government housing and land development programmes (Department of Land Affairs 1997). While there have been some notable cases and successful claims, (for example the 1998 Makuleke claim on the Pafuri region, Kruger National Park), there are currently huge backlogs of cases in all provinces.

In North West Province (which incorporates the former Bophuthatswana) there are over 6,000 cases still to be heard by the Land Claims Court. Two interesting cases (one resolved and one still under consideration) illustrate how this policy in South Africa has direct and indirect impacts upon rural livelihoods. The first example is a claim by the Baphalane people on Madikwe, located close to the Botswana border. The Madikwe land claim is an interesting case of community versus government. The Baphalane came to Sesobe in the 1880s and settled as tenants of the Roman Catholic Mission but were removed from this land during the 1950s, and as tenants, lost their rights to land as soon as they were removed. Today the land in question forms part of the Madikwe Game Reserve (see Box 3.1). While the validity of the Baphalane claim is being investigated by the Land Claims Commission, other considerations also have to be recognized in this case.

Considerable investments have been made in the park—a flagship tourist development in the province—which would be lost should the claim go through, with possible damage claims for breach of agreements with the private sector running into millions of rand. No conclusion has been reached by the Land Claims Commission as to whether the community had ownership rights on the land and this case for restitution remains unresolved. This situation has become highly sensitive: land under claim in the game reserve may not be developed in any way until the claim is resolved, and thus development plans within the park have been seriously delayed. The case of Madikwe is further complicated by considerable donor government aid for example, from DFID (UK Department for International Development) who

Box 3.1. Tourism potential in Madikwe Game Reserve,
          North West Province

The Madikwe game reserve, which was established in 1991 on 60,000 ha
of land in the former homeland of Boputhatswana, is one of the largest
national parks in South Africa. Once fully stocked the park will support
more than 31,000 head of game including the 'big five', the outcome
of the world's largest translocation programme from other southern
African parks. The farmland site which was originally expropriated by
the South African government for redistribution to black cattle farmers
was designated as a national park following commissioned consultations
which concluded that the land, which was heavily bush encroached,
would be more economically and environmentally viable as a game
reserve. With the dissolution of Boputhatswana in 1994, the North West
Parks Board took control of the venture in what they describe as a 'partnership in conservation' between the Board, the private sector, and the
local community. The Parks Board is responsible for conservation within
the park, private-sector investors for the development of game lodges
with the local community benefiting through the creation of employment opportunities (e.g. bush clearance within the park, domestic staff
in lodges), and separate development initiatives in communities away
from reserve. For example, the North West Parks Board was instrumental in securing funding, principally from the Reconstruction and
Development Programme (RDP), for a primary school for the Molatele
community adjacent to the Madikwe park. In addition local community
empowerment is envisaged through the creation of community development projects identified by the communities themselves. To this end, the
Madikwe Community Liaison Forum was established to facilitate this
process. The UK Department for International Development (DFID) is
involved in funding community training for these projects with the aim
that in the future these projects will be financed through an annual divi-
dend paid by the Park. The potential of the venture to improve local
livelihoods in this economically depressed area is contingent upon
private sector funding filtering down to local communities. To date, two
lodges have been built on the reserve employing fifty-seven local people.
Plans to build further corporate lodges in the park on the South
African/Botswana border have been held up by the Department of Land
Affairs refusal to sign leasehold agreements pending the outcome of land
claims in the area. While most of the visitors to date have been from
South Africa, the Parks Board are now targeting the eco-tourism market
by emphasizing the environmental and local community benefits from
tourism in the park. For example the UK-based independent travel

Box 3.1. *Continued*

magazine *Wanderlust* (April/May 2000) recently included a piece on Madikwe selling it 'as the first wildlife park in the world founded purely as the best way to make money for the local community. The local people—some 25,000 people who live in the villages around the reserve—even have a direct say in how the park is run and receive a proportion of the income from tourism. No people were removed to make way for the animals nor have their livelihoods been destroyed or displaced' (p. 38). The article makes no mention of the land claim that is affecting the economic viability of the park nor of the fact that, unlike other examples of community involvement in wildlife management, local communities are denied access to the natural resource base of the park to *improve* their livelihoods.

are supporting the establishment of community projects within the park. While community empowerment and community 'ownership' of the reserve are ultimate aims, it seems initially at least that little power and authority has been devolved. The land claim receives little attention in Madikwe documentation and publicity and no link is made between the claim and the communities involved in the 'participation and development process'.

The critical question here is which option, community or park development, offers the best to surrounding communities in terms of viable sustainable livelihoods. Without the land claim the communities will continue to be involved in the park but whether they will have any real control and authority is questionable. If the claim is awarded, the ownership of the land would be transferred to the community, and they would have far greater control over what happens in the park in terms of decision making, planning, and future developments. It is unlikely that the community would return to the land to resettle given the economic opportunities afforded to them through running a community game park alongside, or as part of, Madikwe. However, this would be a radical and untried change of approach for the Parks Board.

The second case, the land claim on Mosita by the residents of Logageng (Molopo District), illustrates a different set of issues from the Madikwe case. This case is much lower in profile with clear evidence that residents of Logageng once lived in Mosita. Their case has been upheld and in 1999 the claim was awarded. Mosita was a black Native Reserve in an area surrounded by white commercial farms. During the 1960s efforts were made to remove people from the area, and finally in 1968 the people of Mosita were relocated to Logageng, 32 km north, which was the main black Native Reserve. For the last thirty years Mosita has been used by the South Africa Defence Force for training.

Logageng, like most compensatory areas, now has a developed infrastructure and government initiatives have continued to provide services such as schools, clinics, and roads. The settlement has increased in size as people have settled there from the surrounding farms and established families have grown. What is interesting about this case are the different and diverse opinions regarding the land claim that have been revealed through the land claim process. While there is no doubt that the community as whole are celebrating their successful land claim, older members of the community who were initially pushing the claim, now report that they are unlikely to actually return to Mosita. Instead it is their children, the younger generation, who are likely to be the ones resettling in Mosita. This is illustrated clearly in the case study example presented in Box 3.2.

The rationale behind this scenario is that those who moved from Mosita in 1968 have built houses, cleared fields for cultivation, and invested in the land in Logageng. These people say they do not have the energy or inclina-

---

Box 3.2. From Mosita to Logageng, the livelihood profile of Mr Mosingwe, North West Province

---

Mr Mosingwe was born in 1931 in Mosita. As a young man he worked in the mines but returned to farm the land in 1968 when his family and relatives were moved from Mosita to Logageng. He has been farming in Logageng for twenty-three years. When he first started farming he had just a few hectares and some livestock, but when he came to Logageng there was not enough grazing so he sold some of his animals. Now he grows mainly maize and groundnuts and sells most of the crops to the co-operatives. He has just ten cows and his few sheep and goats have died. He has gradually accumulated more land and share-crops with other land owners to increase his access to land. He has tried to set up a small shop in the village but the business has failed, he is not sure why. He also receives a pension and this helps considerably with monthly costs.

He is glad that Mosita has been returned to them: 'It is our land and we love it,' he said. But he says he will remain in Logageng: 'I have put hard work into my fields and I have built myself a house.' 'It will mostly be the younger people who go there,' he says, referring to his children, 'and I will support them with tractors and such things.' There are few livelihood options available to his children if they stay in Logageng but in Mosita they will have access to land, water, and grazing.

tion to do this again. However, they report a shortage of land (both arable and grazing) for younger generations who want to start farming. There are few alternative livelihood options available within these settlements and thus access to land is critical. Thus the younger generation are willing to move to Mosita to build new houses, clear fields, and establish themselves in a new settlement. Logageng will still retain close links with Mosita, through extended families, and visits to ancestral sites, and the two villages will also share a chief. In terms of livelihoods, the restitution of Mosita is providing a new generation with new livelihood opportunities and at the same time re-establishing cultural and emotional links for others. However, the long-term viability of small-scale agriculture in these settlements is another livelihood question to be addressed. Despite agricultural reform in South Africa, there now prevails a controversial debate over small-scale versus large-scale farming in terms of overall productivity, efficiency, sustainability, and long-term contribution to local, regional, and national economies. The White Paper on Agriculture (1995) lays out a clear agricultural strategy based on equity, market forces, and environmental sustainability. Several key issues stand out suggesting the government is serious in its intentions to reform South African agriculture. It aims to create an enabling environment for small-scale farmers, allowing them to contribute to national food security without ignoring the valuable role of large-scale commercial farming within the national economy. To do this requires secure access to land for many of these farmers, and this is addressed through the land reform policy. Small farmers also need access to credit. Formerly credit has been geared to large-scale farming and has not been flexible enough to cope with the diversity and variability of small-scale production, often in more marginal agricultural areas. The government also recognizes that a significant amount of training in how to manage loans is needed at the grass-roots level. Overarching this new agricultural approach is the need to promote environmentally sensitive and sustainable agricultural practices.

Resettlement in Namibia

Unlike that of South Africa, the Namibian government decided that restitution of land to specific ethnic groups was not feasible as the multitude of overlapping rights claimed by numerous groups on any one piece of land over decades and centuries was seen as too complex to resolve equitably and peacefully. There were also fears that to hand over commercial farmland would have adverse affects on economic productivity; the World Bank (1991) has, for example, argued that in Namibia 'efficient land use and environmental protection are probably incompatible with a radical redistribution of land from large-scale to small-scale farmers' (quoted in Pankhurst 1995: 555). Thus the Namibian Directorate of Resettlement and Rehabilitation opted for a policy to resettle the landless and displaced and to alleviate overcrowding in communal areas without disturbing the status quo. Most resettlement was therefore focused on less-crowded communal areas. There are currently forty-one resettlement projects across Namibia, with a population of more than 15,000 people resettled in communal areas and about 2,000 people resettled on twenty-eight farms in commercial areas.

The mission statement for the Division of Resettlement is premised on the notion that the government has a particular responsibility to uplift the living standards of all its citizens. Once again livelihoods are explicitly linked with land reform. Its main objectives are to redress the past imbalances in the distribution of natural resources, particularly land; to provide opportunities for the disadvantaged to produce their own food and to sell the surpluses to the market and by doing so contribute to national food security; to give opportunities to the landless to own land; to bring small-holder farmers into the mainstream of the Namibian economy; to create employment through full-time farming; and to help in the process of poverty eradication (MLRR 1999).

Despite the reporting of many seemingly successful programmes (see ibid.), there is evidence to show that resettlement has at times been no more than simply relocation. The plight of the Seeis squatters, for example, has been highlighted in a government publication as 'an issue of grave concern since the dawning of Namibia's independence' (ibid. 30). Most of the squatters were retrenched farm-workers who had been made landless and homeless by previous employers and had nowhere to go. The government of Namibia was taken to court by the farm owners and in late 1997 the resettlement of the Seeis squatters commenced. A total of 115 families were 'successfully resettled' on a farm in the Omaheke Region on 4,315 ha of agricultural land comprising thirty-six camps. However the case illustrated in Box 3.3 suggests that the resettlement has not yet been successful.

The policy promotes two modes of resettlement, individual plot holding or a collective undertaking. Prospective settlers have the right to choose the way they want to settle and where they want to settle. However, in order for resettlement to take place the government must purchase land, and this is the major obstacle to the land reform process. Land acquisition under this policy requires a 'willing seller willing buyer' situation (ibid. 16). This means the government can only acquire land provided that there is an offer on the market. Very often significant time is spent on price negotiations after which the two parties (the government and the willing seller) fail to reach an agreement (ibid.). Thus only a fraction of cases result in land acquisition with fewer choices open to those being resettled.

The implications for livelihoods is significant, as clearly illustrated in the case of Varsdrai, because there are often few livelihood opportunities available to those being resettled. Although those resettled have more 'freedom' to keep livestock, this is of little consequence when grazing conditions are poor and areas are overstocked and overpopulated. The measures in

Box 3.3. Resettlement at Varsdrai, Omaheke District

Isaac has been living in Varsdrai for two years. He used to work at various white farms but ended up at Seeis. He was there for a few months or a year, he is not sure. There were too many people at Seeis and he was brought to Varsdrai by the government. He now lives in a small paddock which he shares with three other households. Each has its own goats but cannot graze them outside the paddock, this land has been allocated to other people who have their own livestock. 'We do not get help here . . . each person suffers on their own. We used to stay in a long line with better housing, we are not used to these small paddocks. The problem is there are too many livestock for the area,' reports Isaac. Isaac's neighbour Margaret points out the disillusionment that many residents report, 'We were told we should have our own gardens and live from them but we have water problems. The borehole does not have water. We were told a project would open for old people, gardening and farming. It was going to be a government project and they were going to send people, but it is still quiet . . . When we were still at Seeis we were told by the government that they would help us. Now there are no jobs, nothing.'

operation to assist the establishment of new communities are clearly weak and external attitudes to those who have been relocated reinforce images of destitution and dependency.

Resettlement in Botswana

Post-colonial policy discourses concerning access and entitlement to land in Botswana, untainted by the legacy of apartheid, have emphasized resettlement rather than redistribution and restitution principally through the Remote Area Development Programme (RADP). The RADP was set up in 1977, formerly the Bushman Development Programme, to address rural poverty by targeting populations who lived outside village settlements and destitutes within villages who were socially and economically marginalized (Hitchcock 1997). This policy has been widely associated with Basarwa groups, who were initially targeted in 1974 under the Bushman Development Programme, and has been criticized for failing to recognize the diversity of needs among Remote Area Dwellers (RADs) and, in particular, between different Basarwa groups (Hitchcock 1996). A key component of the RADP was to resettle remote area dwellers into communal service centres where access to employment opportunities and health and educational facilities could be centralized. The intention was that each service centre should be established around an existing, reliable water source in an area with sufficient land for grazing, hunting, and gathering with fertile soil for crop and vegetable production. In addition to livestock and smallstock keeping, other agricultural activities and employment projects, such as carpentry, sewing, and leatherwork were established to provide the basis for sustainable income generation. While livestock and smallstock projects have been enthusiastically received by local populations, the introduction of other income generation projects has suffered due to lack of training, materials, and access to markets (Chr. Michelsen Inst. 1995). Service centres are also the loci for labour-based drought relief, feeding programmes, and labour-based public works programmes providing a source of cash income in exchange for labour.

The expansion of commercial pastoralism into the Kalahari through the Tribal Grazing Land Policy (see Ch. 6) and the progressive privatization of communal lands through fencing has placed increasing pressure on these service centres to absorb displaced populations. Today, service centres are characterized by high levels of unemployment and welfare dependency such that many RADs are being forced back onto farms and cattleposts where they exist either as squatters or are exploited as a reserve of unwaged labour (Sporton *et al.* 1999). According to Hitchcock and Holm (1993) the only thing keeping people at service centres is the drought relief that is distributed at these settlements, for example, between 1982 and 1990, up to 90 per cent of all RADs were dependent on food relief.

The case study of Khawa RAD Service Centre (Box 3.4) highlights key aspects of the failure of service centres to build rural capacity and promote sustainable livelihoods. While the service centre was established at a site of existing population settlement, there was little consultation and planning with, in particular, the absence of potable water and poor communications affecting the physical capital component of livelihoods.

Box 3.4. Khawa Service Centre, Kgalagadi District,
south-west Kalahari

Khawa is the main service centre for the RAD population of the south-west Kalahari who have been displaced by the establishment of commercial ranches and farms in the area. The settlement has no potable water supply and is reliant on deliveries of water every two weeks. There is a primary school and clinic; however, mobile clinic services are sporadic due to the remoteness of Khawa and lack of investment in vehicles. Several income-generating projects have been established including a bakery, a co-operative store, carpentry, bee-keeping, and

Box 3.4. *Continued*

sewing. In 1993 the RAD Office gave twenty-five households ten goats from which they were supposed to breed, a scheme which had limited success as many died in the drought or were slaughtered for consumption. Ten households who had successfully bred their goats were given five cattle in 1999. Although villagers were given beehives and flower seeds to plant to encourage bees, most of the bees disappeared with drought. Of the other projects, the bakery has failed because of lack of investment, the sewing project survives making school uniforms, and the carpentry project and co-operative are still in operation. Despite the introduction of these projects there is general agreement that they benefit only a limited number of people and the livelihoods of the majority are dependent on labour-based drought-relief projects and destitute food rations and payments. For example, one senior Remote Area Development Officer commented on training projects; 'It's just a waste. If the government could ensure that after a period of training these people received jobs, things would be a lot better. At the moment, people don't want to bother with the skills training. They see others who have undergone training and still have no work. They think it is pointless to learn these skills. The solution to many of this district's problems is the creation of more jobs.' This situation has been exacerbated in recent years by the in-migration of destitutes from elsewhere in search of RAD benefits such as the payment of school expenses for those who live in the settlement. Failure to provide employment has resulted in a surplus workforce who move between Khawa and local farms in search of work or handouts. John, a Mokgalagadi man in his forties, now resides on a commercial ranch where he works for food rations from the ranch owner. 'Before I was just staying at Khawa doing nothing. There is no work for a person there at Khawa. Before I did the drought relief. You do labouring on the drought relief. Making those concrete houses for the nurses and teachers. The work was hard but okay. Then the work was finished. There is not enough work for all the people of Khawa. I asked the village council for help. They told me I am fit to work and must look for a job. They only want to help the Basarwas. Not the Bakgalagadi. If the [ranch] owner says I can stay and then pays me then I would like to stay here. Maybe not for ever. But for a long while. If not then maybe I will go back to Khawa. maybe the drought relief will take me again. I could do labouring jobs for the drought relief . . . I would never leave this place, the Kalahari. I have always lived in this area. It is my home. No, if I leave here I will return to Khawa. I am too old to go far away from this area to find a job.'

Many have been drawn to Khawa by the prospect of assistance through various safety-net policies such as RAD school aid and drought relief resulting in overpopulation, restricted grazing, and the depletion of natural resources. The top-down RAD policy has backfired—instead of providing viable livelihoods for those in Khawa it has created a surplus population or underclass dependent on the state. If the ongoing privatization of communal lands continues then more people will be displaced to such centres making the livelihood transition from self-sufficiency to dependency. Received wisdom in policy circles states that the privatization and commercialization of communal areas in favour of the few will enhance the national economy, encourage environmentally sustainable practices, and ultimately benefit all. Lessons drawn both from within Botswana (e.g. Thomas and Sporton 1997) and from elsewhere suggest that common property regimes are vital to the livelihood strategies of the poor in marginal environments.

The Commercialization of Agriculture

Within the Kalahari, agriculture encompasses both extensive and intensive arable and pastoral farming practices. The agriculture sector largely remains dualistic, with a relatively small number of large commercial farms and a large number of diverse smallholder farms. Typically the commercial farms occupy the most favourable areas, are highly mechanized, and have enjoyed subsidized inputs (particularly in apartheid and colonial eras). Conversely smallholders typically occupy communal areas in more fragile and marginal environments, with poor extension services and reduced levels of subsidies (Whiteside 1998). Botswana differs slightly from Namibia and South Africa as it has a small-scale but well-established indigenous commercial farming base.

Spatial variations in rainfall mean that the suitability of different areas of southern Africa for crop production varies enormously. Population densities within individual countries are very variable and can reflect historical factors as well as agriculture productivity. Critically, even within countries with low overall population densities, unequal distribution of land and people means that there are areas of acute land pressure (ibid.). Commercial and subsistence livestock production face similar problems with rainfall again being the most important limiting factor. In some communal areas pressure on rangeland resources has resulted in compositional change to the natural resource base (e.g. from perennial to annual grasses) which causes acute problems in low rainfall years, but does not necessarily affect overall productivity in average (or good) rainfall years (White 1993). Recent severe droughts have undermined smallholder resources across southern Africa and left many of the poorer households without cattle. This has had knock-on effects on draught-power, land cultivation, and fertility maintenance. Policies for small-holder support have been largely founded on the transfer of technology approach although recently, more holistic and participatory approaches have been incorporated echoing a wider shift across policy spheres towards these types of approach.

What is clear in southern African agricultural policy is the move to com-mercialize small-scale production, open up markets, and integrate more effec-tively (black) rural areas into the national economies of the countries. Two cases that illustrate these shifts quite clearly are the moves to privatize com-munal rangelands in both Namibia and Botswana, and the current financial dilemmas facing small-scale arable and mixed farming in South Africa and Botswana.

Enclosure of the Range

Enclosure of the land through private fencing is one of the major contem-porary problems in the communal areas of Namibia, where the majority of the rural population lives. Fencing in communal areas started in the mid-1980s placing great pressure on adjacent areas where communal farmers are being forced to relocate. Further studies suggest that fencing of the commu-nal lands is being done by 'those with money'. However, in some areas local people also suggest that political patronage and close relationships to local political élites are prerequisites for erecting a fence. In other areas, illegal fencing has been so pervasive that for some households erecting fences has become a source of cash income. Furthermore, some communities have retaliated with 'defensive fencing' where fences have been erected to prevent traditional grazing areas from being enclosed by people from outside the community. Semi-commercial farmers who erect fences are obtaining exclu-sive rights of access to rangeland resources and are able to utilize dual grazing rights on the remaining communal land. As most of these enclosed areas include watering pans, communal farmers are also denied access to water (Fuller and Nghikembua 1996). Such activities can displace existing small-scale communal farmers, thus placing added pressure on lands surrounding communal areas, or force them to relocate to other already overcrowded com-munal lands or to areas reserved for other activities (e.g. wildlife).

Section 38 of the Agricultural (Commercial) Land Reform Act of 1995 has legitimated this process by making provision for the subdivision of land acquired under this Act into surveyed holdings for small-scale farming pur-poses, thereby constraining potential forms of communal land and resource management. Interestingly, these restrictions echo conditions laid down for European settler farmers by the colonial authorities (Sullivan 1999). In other words, the Act advocates the further enclosure and subdivision of land through fencing. The draft Communal Lands Reform Bill also places empha-sis on the subdivision of communal land into alienated land holdings. It proposes that any person or group of persons holding recognized rights to communal land is 'entitled to convert such holding into a leasehold tenure of one hundred years' providing this takes into consideration local customary law. Similarly, vacant communal land may be delineated and allocated as economic land units. Allocation and management will ultimately rest with Regional Boards (modelled on Land Boards in Botswana). These will have the power to allocate rights under customary law in communal lands falling under its jurisdiction, to cancel these rights, to allocate land, and to demarcate land into economic holdings. A Land Adjudication Commission will be established to mediate disputes. The bill is still under development and has already gone through many changes (R. Rohde, pers. comm. 1999). Furthermore, given the controversy surrounding the bill it is unlikely that it will be put through parliament before the election in 2000 (J. Barns, pers. comm. 1999).

Despite the new reforms, the policy remains on track to transform traditional stock-keepers into commercial farmers and replace customary forms of land tenure with freehold or leasehold title through the subdivision of communal land into fenced holdings. Both policies promote the privatization and further enclosure of commercial and communal lands, and have led in some communal areas to land grabbing and defensive fencing. Such exclusion of people and livestock from 'former' communal lands exacerbates pressure on resources on the remaining communal lands and promotes the chaotic management systems that prevail.

Mobility is a fundamental component of herd management in arid and unpredictable environments (Scoones 1995; Sullivan, 1999). Even when people have access to delineated areas of land this implies that the current policy emphasis on subdivision of land into individual land holdings may be inappropriate in some of Namibia's communal areas (Sullivan 1999: 147). Furthermore, the World Conference on Agrarian Reform and Rural Development (WCARRD) Review Mission, funded by FAO (1993), stated that there should be a moratorium on fencing communal rangeland, and that communal areas should retain communal systems of land tenure, on the understanding that these systems best ensure security of tenure for the rural poor. The knock-on effect of fencing on livelihoods in the communal areas needs to be addressed. Box 3.5 describes how one man has benefited from the increase in fencing in the area. However, he is also aware of some of the problems it creates. Such cases illustrate how policy directives are manipulated and reified on the ground through both positive and negative actions, and these in turn have far-reaching effects on livelihood sustainability and household vulnerability.

In Botswana, an important policy for the livestock industry was the 1975 Tribal Grazing Land Policy (TGLP). There were three goals to the TGLP: to improve range management and prevent overgrazing and further environmental degradation; to bring about greater equality of rural income; and to

Box 3.5. Enclosure in Omaheke South Communal Areas, Namibia

Eben Kandetu is 38 years old, the eldest son of a Herero family originally from Epikuro. He now lives in Springwater, a settlement well known recently for its controversial fencing. His family came to Springwater five years ago in search of better grazing because of drought: most of their cattle had died. Eben now makes a living through erecting fences for people both in Springwater and elsewhere. He says the paddocks are usually just for breeding but sizes range from 7 by 7 km to 10 by 14 km. 'There are big paddocks around here. People hit each other for these big paddocks. It is just some people, those people who have money,' he reports. 'People with money put up fences and don't consider those people who don't have money,' he continues. He says that the grazing is good in the area but there is not enough and so some people have started to move away. 'The only grass around here is sour grass [an annual species] . . . The paddocks do not really help, it just stops my livestock from straying . . . But when the grass is finished I will just put them in the paddock,' he explains. Eben says that there are only about six households left in Springwater. Many people have left because of the arguments about the livestock and fencing.

foster growth and commercialization of the livestock industry (for detail, see Ch. 6). Land was zoned into three categories: commercial, communal, and reserve. Areas zoned for commercial ranches assumed to be uninhabited were in fact occupied by a significant number of people. Wildlife Management Areas were then proposed for the reserved land to accommodate people adversely affected by TGLP (see Ch. 7). Severe problems emerged with TGLP in relation to the use of dual grazing rights in the communal lands leading to increased pressure on the range and little incentive for 'good' management practices in ranches (Peters 1995).

The failure of the TGLP to achieve its goals led the government to reconsider its agricultural policy in the 1980s, culminating in the 1990 National Policy on Agricultural Development. Limited land resources for livestock expansion and a change in the global economic climate indicated to the government that Botswana's expansionist approach to increasing livestock returns needed to be reconsidered. The new policy promotes the intensive use and efficient management of land resources and livestock to achieve its goal of increased productivity in this sector. One means of achieving this goal is through the fencing of some of the communal land areas. This is also seen as a step towards addressing the environmental and economic problems associated with or emanating from the communal grazing management system. Poor management has been identified as the most significant factor contributing to the poor performance of the agricultural sector (Balopi 1996).

The potential impact of this policy on rural livelihoods in communal areas is critical. Experiences from other countries in Africa, and elsewhere around the world, suggest that privatization of the communal lands can have devastating effects on the sustainability of rural livelihoods for the rural poor. It is also uncertain what overall environmental benefits result from these changes as grazing intensities could be concentrated further as the poor or marginalized people are fenced out. The impact fencing could have on wildlife in rural areas is also another crucial issue with implications for biodiversity, poverty alleviation, and sustainable livelihoods. Decline of wildlife populations in Controlled Hunting Areas could potentially put further pressure on resources in Wildlife Management Areas either through violation of hunting regulations or political pressure to free up areas within the Wildlife Management Areas to citizen hunting. There is also increasing pressure to dezone land uses such as Wildlife Management Areas and Game Reserves to give way to cattle grazing (ibid.). Further, with the privatization of the communal lands, people will inevitably be displaced (Machacha 1996; Balopi 1996; Maroba 1996) and will almost certainly turn to the Wildlife Management Area settlements as places to settle and gain access to alternative water and grazing rights. This puts further pressure on already limited resources as Wildlife Management Areas are situated in what the government describes vaguely as 'marginal' lands.

Botswana is facing increasing pressure to enclose, or privatize, its communal grazing areas. Alongside this, there are also pressures to introduce community-based natural resource management projects in areas such as Wildlife Management Areas (Ch. 7). These two policies do not necessarily sit easily side by side and Botswana will have to resolve policy conflicts across these related sectors if it is to achieve social justice, ecological sustainability, and economic stability at the local, regional, and national levels.

Small-Scale Farming

In the Kalahari regions of South Africa, a dual system of agriculture emerged through apartheid rule with white large-scale farming heavily subsidized by the government and black small-scale farming confined to the subsistence sector in homeland areas which were marginal in terms of agricultural productivity. Despite agricultural reform in South Africa, there now prevails a controversial debate over small-scale versus large-scale farming in terms of overall productivity, efficiency, sustainability, and long-term contribution to local, regional, and national economies. Given the critical importance of agriculture to South Africa's economy, the agricultural policy reform process is likely to be the key policy shaping rural livelihoods in South Africa in the years to come.

The White Paper on Agriculture (1995) lays out a clear agricultural strategy based on equity, market forces, and environmental sustainability. Several key issues stand out suggesting that the government is serious in its intentions to reform South African agriculture. It aims to create an enabling environment for small-scale farmers, allowing them to contribute to national food security without ignoring the valuable role of large-scale commercial farming within the national economy. To do this requires secure access to land for many of these farmers, and this is addressed through the Land Reform Policy. Small farmers also need access to credit. Formerly credit has been geared to large-scale farming and has not been flexible enough to cope with the diversity and variability of small-scale production, often in more marginal agricultural areas. The government also recognizes that a significant amount of training in how to manage loans is needed at the grassroots level. Overarching this new agricultural approach is the need to promote environmentally sensitive and sustainable agricultural practices.

The South African government has had five years in which to start implementing and reforming agriculture in the rural areas. However in many rural areas few farmers have seen the changes they expected. In North-West Province it is clear that much of the new agriculture policy is yet to trickle through. While access to land in this Province is secure, and in places has been enhanced through the award of land claims, many farmers are still facing considerable hardship and the question 'where is the life in farming?' is often heard. While many of their problems and difficulties have been identified in the agriculture policy and White Paper, few have actually been addressed on the ground. This highlights one of the key constraints, that the policy is not being translated into practice and thus rural livelihoods are not being enhanced. Box 3.6 illustrates the case of Mr Maele, a semi-commercial small-scale farmer, and the constraints that he, and others, face in making a living through cultivation in rural South Africa.

Agriculture in Botswana can be divided into two distinct sectors: traditional and commercial. Commercial agriculture is largely confined to the freehold and leasehold farms, whereas traditional farming is exclusive to the communal areas. The commercial sector in Botswana uses modern technology and applies purchased inputs such as hybrid seeds, agrochemicals, and exotic livestock breeds. It is also more likely to require hired labour, while traditional agriculture typically uses family labour. Commercial agriculture is integrated in formal markets, but the traditional sector produces mainly for its own consumption, only selling when surpluses are available. The traditional sector also includes resource-poor farmers who do not own cattle, are engaged in marginal crop production, and are net food buyers. Such households supplement their own production through wage earnings, transfers, and off-farm activities. Commercial farms comprise 8 per cent of the total land area and tend to specialize in cattle production while traditional farms cover about 70 per cent of Botswana's total land area, a considerable contrast to

Box 3.6. Mr Maele, a small-scale semi-commercial farmer in North West Province, South Africa

Tiro Maele is a farmer in Mathateng. He lives with his wife, three adult children, and eight children and grandchildren. One of his daughters works as a cleaner in a garage in Gauteng and she brings home money each month to the family. Tiro was unable to cultivate in 1999 because of money problems. He usually plants about 29 ha with maize and groundnuts. He has used chemical fertilizer in the past but not recently because of a shortage of money. Tiro explained, 'I borrowed money from the Land Bank in Pretoria in 1994. I filled in the forms in Mafikeng and they sent them to Pretoria. I borrowed R4,000 but the crops failed and I could not pay it back . . . I used the money to repair the tractor.' Though Tiro has sold some of his cattle to pay off the loan he still owes some money. Until he pays off the remainder of the loan he cannot reapply. However, in order to plough he must have money for diesel, seeds, and, if possible, fertilizer. 'I want to convert back to donkeys and plough with them because it is cheaper on fuel costs,' he explains. However, it is not easy to buy donkeys locally. In 1983 donkeys in the former Bophuthatswana were killed as part of a strategy to combat degradation and in an attempt to force small-scale farmers to mechanize. Donkey plough teams were slaughtered and only a few donkeys remained for transport. Most farmers bought tractors at this time and donkey populations have not recovered.

the situation in Namibia. Two-thirds of traditional farmers practise mixed farming, with individual management of arable holdings and communal grazing livestock (Republic of Botswana 1997).

The Barolong farms in south-east Botswana were at one time regarded as the bread basket of Botswana with both commercial and traditional arable production operating side by side. However, within the last decade there has been a significant decline in the amount of land under cultivation and overall yields for the areas have dropped significantly. There are several reasons that may account for this decline. Within the last decade Botswana has shifted from a policy based on food self-sufficiency to one of food security. Botswana now freely imports much of its grain from South Africa and as a consequence gives less support in subsidies and overall encouragement to its own farmers in dryland areas. These farmers cannot compete with the South African market and thus less land is being cultivated under traditional crops (white maize and sorghum) and farmers are starting to diversify away from cereal production.

Small-scale farmers in the Barolongs faced further problems when the National Development Bank (NDB) recalled agricultural loans taken out in the 1980s to buy farm machinery. These loans had mainly been used to mechanize small-scale production in Botswana at a time when Barolong agriculture was being strongly promoted. However, with the change in economic climate in the mid-1990s, the NDB recalled their loans in 1995 and where necessary confiscated tractors and other farm machinery from farmers defaulting on their loans. Payments on many loans had ceased following drought in 1992 and again in 1994. Most farmers had not kept up their donkey or oxen plough teams and thus most farmers were unable to plough after the confiscation of their tractors. This had severe effects upon their agricultural livelihoods and has possibly led to greater reliance on state support systems and remittance payments from nearby towns and cities.

The change in focus of the Agriculture Policy, combined for many with the loss of farm machinery, has considerably increased farmers' vulnerability to shocks and stresses. Drought is endemic to the region and both major and minor drought incidences have had severe impacts upon small-scale farming livelihoods. The cases outlined in Box 3.7 show how three very different farmers have been affected by changes in policy and how their differing levels of vulnerability make them more or less susceptible to variability in the environment.

Botswana's food requirements are increasingly purchased by export earnings from the diamond sector. However beef export earnings still adequately cover imports of basic cereals. Productivity in the livestock sector has been hampered by persistent drought, which has resulted in shortage of water and poor grazing conditions. Productivity in the arable sector is still low. Here the

---

Box 3.7. Farmer case studies from the Barolongs, Botswana

Mr Abraham Molophodi: Abraham has access to over 700 ha of land in and around Mokatako. His main occupation is farming, though he is also councillor for the area. He runs a small shop in the village and his wife is a teacher in the local primary school. Abraham started ploughing in 1983 with 70 ha. A year later he bought his first tractor with a loan from the NDB. Since then he has expanded every year and now has a fleet of eight tractors, over 400 ha of his own land, and access to 300 ha more through share-cropping agreements. Alongside his successful farming he keeps 350 cattle and over 150 goats and sheep. However, he finds the financial condition of farming in Botswana increasingly harsh with input prices rising while output prices are falling. Abraham has good access to credit and is considering diversifying away from white

Box 3.7. *Continued*

---

maize and sorghum into yellow maize which will be sold fresh to the nearby market in Lobatse. He hopes this crop will command better prices and raise the profile of his farming once again.

Mr and Mrs Mopako: Michael and Kenelwe have a 15 ha plot in Mokatako which they acquired in 1986. Michael says he does not borrow any extra land because he ploughs with donkeys and they are very slow. Both have vegetable gardens which they water, one by the house and another by the field. Through an agricultural scheme Michael has built a rainwater storage tank on his farm which he can use to water his donkeys during ploughing. This means his ploughing is more efficient and he can take more advantage of rains when they come. 'In the past I have harvested 200 bags from the field but now there is no money so even if it does rain there is no money for fertilizer,' he says. They usually plant a variety of crops on their field: yellow and white maize, two types of sorghum (phohu and mahube), and several types of beans. He gets some seeds free from the agricultural extension officer which saves him money and allows him to experiment with different varieties. Because of the recent poor harvests and the lack of grazing in the surrounding areas for his cattle, he is considering planting crops specifically for cattle-feed next year. He is currently financing his farming through livestock sales.

Ms Maria Modisaotsile: Maria was born in 1953 in Mokatako where she lives now. She has never been married and has been farming since 1979: she has always farmed on her own. She likes to plough the whole of her 10 ha field but this is not always possible. Her harvests have been poor recently and she attributes this to unreliable rainfall and late ploughing. 'In my life I have never done any other job except ploughing. It is very difficult. I do this job and work hard,' she explains. She did not plough this year because of the lack of rains and she was also sick; last year she only managed to plant 2 ha because she was sick. She planted maize and sorghum but did not harvest anything, 'the birds took it all'. To survive her children have been sending her money this year. She has saved some of the money meant for food to buy a donkey cart. She also wants to save to buy a new plough. She has donkeys which she bought with the help of the government, they also helped her with wire to fence her field, but not with the poles. Through the agricultural schemes the government will also help her to buy a plough. She also lost her cattle this year. She has registered them with the police as missing but it is unlikely that they will be returned. She now has just six donkeys for ploughing and two goats.

protracted incidence of poor weather conditions characterized by variable and low rainfall have adversely affected productivity in this sector. However, of further concern to the Botswana government is the incidence of HIV/AIDS-related diseases and their impact on the performance of the agricultural sector (Republic of Botswana 1997). These diseases may have adverse effects on the provision of labour for agricultural production and off-farm employment and this may in turn negatively affect household economic situations and food security. The impact of HIV/AIDS-related diseases on subsistence (as well as formal) labour provision is also a serious concern for South Africa and Namibia.

Concluding Comments

Under colonialism and apartheid various restrictions were placed on land rights and land use. The challenge of tenure reform now is to de-racialize the system of land rights in a way that brings pre-existing vested rights in land within a non-racial unitary system. The challenge is to overcome current tenure disputes, overlapping tenure rights, and conflicting land claims. New policies *must* broaden the base of land ownership. Evidence from land reform around the world has demonstrated that many farm operations on smaller scales can be more efficient and that common property regimes can be viable, but such reforms must be balanced with overall food security concerns at local, regional, and national levels. Central to policies should be recognition of the need simultaneously to promote sustainable livelihoods. While the three countries here illustrate that policy and livelihood creation and enhancement are critically linked, all need to address more radically issues of empowerment, self-sufficiency, and diversity within rural livelihoods. While South Africa has taken the bottom-up approach to policy, the implementation of both land and agricultural policies has been delayed, partly through consultation, such that some commentators believe that the demand for land and secure livelihoods through rapid rural population growth will now outstrip the current capacity of policy to meet such demand (May *et al.* 2000).

References

BALOPI, P. K. (1996), 'Opening Speech for the Conference on the Fencing Component of the National Policy on Agricultural Development', paper presented at the Conference on the Fencing Component of the National Policy on Agricultural Development, June 1996, Gaborone.

BERNSTEIN, H. (1996), 'South Africa's agrarian question: Extreme and exceptional', in H. Bernstein (ed.), *The Agrarian Question in Southern Africa* (London: Frank Cass), 1–52.

BERNSTEIN, H. (1998), 'Social change and the South African countryside? Land and production, poverty and power', *The Journal of Peasant Studies*, 25/4: 1–32.

CHR. MICHELSON INSTITUTE (1995), *NORAD's Support of the Remote Area Development Program in Botswana: An Evaluation Report* (Bergen: Chr. Michelson Institute for Development Studies and Human Rights).

DEBENHAM, F. (1948), 'Report on the resources of Bechuanaland, Nyasaland, Northern Rhodesia, Tanganyika, Uganda and Kenya', *Colonial Research Publication*, 2: 31–9.

DEPARTMENT OF LAND AFFAIRS (1997), *Thriving Rural Areas: Rural Development Framework* (Pretoria: Department of Land Affairs).

FAO (1993), *Inter-agency MCARRD Policy Review Mission to Namibia* (Rome: FAO).

FULLER, B., and NGHIKEMBUA, S. (with T. F. Irving) (1996), 'The Enclosure of Range Lands in Eastern Oshikoto Region of Namibia', *Social Science Division Research Report, 24* (Windhoek: SSD, Multi-Disciplinary Research Centre, University of Namibia).

GOOD, K. (1993), 'Interpreting the exceptionality of Botswana', *Journal of Modern African Studies*, 30: 69–95.

GORDON, R. (1992), *The Bushman Myth: The Making of a Namibian Underclass* (Oxford: Westview).

HITCHCOCK, R. K. (1985), 'Water, land and livestock: The evolution of tenure and administrative patterns in the grazing areas of Botswana', in L. Picard (ed.), *The Evolution of Modern Botswana* (Nebraska: University of Nebraska Press), 84–121.

——(1996), *Kalahari Communities: Bushmen and the Politics of Environment in Southern Africa* (Copenhagen: International Work Group For Indigenous Affairs).

——(1997), 'Cultural, economic and environmental impacts of tourism among Kalahari bushmen', in E. Chambers (ed.), *Tourism and Culture: An Applied Perspective* (New York: State University of New York Press), 93–128.

HITCHCOCK, R. K., and HOLM, J. D. (1993), 'Bureaucratic domination of hunter-gatherer societies: A study of the San in Botswana', *Development and Change*, 24: 305–38.

JONES, S. P. (1999), ' "To come together for progress": Modernization and nation-building in South Africa's Bantustan Periphery—the case of Boputhatswana', *Journal of Southern African Studies*, 25/4: 570–606.

LAWRENCE, M., and MANSON, A. H. (1994), ' "The dog of the Boers": The rise and fall of Mangope in Boputhatswana', *Journal of Southern African Studies*, 20/3: 447–61.

LEE, R., and DeVORE, I. (1968), *Man the Hunter* (Chicago: Aldine).

MACE, R. (1991), 'Overgrazing overstated', *Nature*, 349: 280–1.

MACHACHA, B. (1996), 'Dual grazing rights', Paper presented at the National Conference on the Fencing Component of the National Agricultural Development Policy, Gaborone June 1996.

MANSON, A. H., and DRUMMOND, J. (1991), 'The evolution and contemporary significance of the Boputhatswana–Botswana border landscape', in D. Rumley and J. V. Minghi (eds.), *The Geography of Border Landscapes* (London: Routledge), 217–42.

MAY, J., ROGERSON, C., and VAUGHAN, A. (2000), 'Livelihoods and Assets', in J. May (ed.), *Poverty and Inequality in South Africa: Meeting the Challenge* (London, ZED Books), 229–58.

MAYENDE, G. P. (1993), 'Bureaucrats, peasants and rural development policy in Botswana', *Africa Development*, 18/4: 57–78.

MINISTRY OF LANDS RESETTLEMENT AND REHABILITATION (1999), *Annual Report 1996/7 and 1997/8 for the Ministry of Lands, Resettlement and Rehabilitation* (Windhoek).

MOROBA, S. S. (1996), 'Compensation issues with regard to the implementation of the fencing component of the National Policy on Agricultural Development', Conference on the Fencing Component of the National Agricultural Development Policy, Gaborone, June 1996.

MOTZAFI-HALLER, P. (1998), 'Beyond textual analysis: Practice, interacting discourses,' and the experience of distinction in Botswana', *Cultural Anthropology*, 13/4: 522–47.

PANKHURST, D. (1995), 'Towards reconciliation of the land issue in Namibia: Identifying the possible, assessing the probable', *Development and Change*, 26: 551–86.

PARSONS, N., and CROWDER, M. (1988) (eds.), *Monarch of All I Survey: Bechuanaland Diaries 1929–37* (Gaborone: Botswana Society).

PETERS, P. E. (1994), *Dividing the Commons: Politics, Policy and Culture in Botswana* (Charlottesville: University of Virginia Press).

PLATZKY, L., and WALKER, C. (1985), *The Surplus People: Forced Removals in South Africa* (Johannesburg: Raven).

REPUBLIC OF BOTSWANA (1997), *National Development Plan 8: 1997/8–2002/03.* (Gaborone: Ministry of Finance and Development Planning).

SCHAPERA, I. (1943), *Native Land Tenure in the Bechuanaland Protectorate* (Alice, Tex.: Lovedale).

SCOONES, I. (ed.) (1995), *Living with Uncertainty: New Directions in Pastoral Development in Africa* (London: Intermediate Technology).

SPORTON, D., THOMAS, D. S. G., and MORRISON, J. (1999), 'Outcomes of social and environmental change in the Kalahari of Botswana: The role of migration', *Journal of Southern African Studies*, 25: 441–59.

SULLIVAN, S. (1999), 'People, plants and practices in drylands: Socio-political and ecological dimensions of resource use by Damara Farmers in North-West Namibia', Ph.D. Thesis, University College, London.

TAPSCOTT, C. (1993), 'National reconciliation, social equity and class formation in independent Namibia', *Journal of Southern African Studies*, 19/1: 29–39.

THOMAS, D. S. G., and SPORTON, D. (1997), 'Understanding the dynamics of social and environmental variability: The impacts of structural land use change on the environment and peoples of the Kalahari, Botswana', *Applied Geography*, 17/1: 11–27.

VAN DER POST, L. (1958), *The Lost World of the Kalahari* (New York: William Morrow).

WHITE. R. (1993), *Livestock Development and Pastoral Production on Communal Rangeland in Botswana* (Gaborone: Botswana Society).

WHITESIDE, M. (1998), *Living Farms: Encouraging Sustainable Smallholders in Southern Africa* (London: Earthscan).

WILMSEN, E. N. (1989), *Land Filled with Flies: A Political Economy of the Kalahari* (Chicago: University of Chicago Press).

WORLD BANK (1991), *Namibia: Poverty Alleviation with Sustainable Growth* (Washington DC: World Bank).

WORLD BANK (2000) (ed.), *World Development Report 2000/2001: Attacking Poverty* (Oxford: Oxford University Press).

WYLIE, E. (1994), 'Hunter-Gatherers in Botswana and the Land Issue', *Indigenous Affairs*, 2: 7–19.

YEAGER, R. (1989), 'Demographic pluralism and ecological crisis in Botswana', *Journal of Development Studies*, 23: 385–404.

4

The Impact of Cattle-Keeping on the Wildlife, Vegetation, and Veld Products

Jeremy S. Perkins, Greg Stuart-Hill, and B. Kgabung

Introduction

Botswana's semi-arid climate, periodic droughts, poor soils, and lack of perennial water supplies, coupled with its Protectorate status of indirect colonial rule by the British, meant that at Independence in 1966 it was one of the least developed countries in the world. Rapid economic growth has ensued, due primarily to the exploitation of diamond reserves in the Kalahari (Morna 1979), but also to foreign aid and a generous EU agreement on beef exports. The latter is commonly termed the Beef Protocol Agreement. It has been severely criticized from some quarters for directing the country along a trajectory of development that is widely regarded as being both environmentally unsustainable and socially and economically damaging, except to a small influential minority of large cattle-owners (Perkins and Ringrose 1996).

A number of interconnected myths continue to direct policy on Botswana's rangelands (Perkins 1996), that remains firmly entrenched in the privatization and commercialization of land through fencing as the way to curb what is often stated to be widespread range degradation. These issues need not be elaborated further here, and while Abel and Blaikie's (1989) criticisms of Botswana's government policy are supported, this chapter argues that increased flexibility in terms of both livestock and wildlife management is the key to resolving the worsening poverty profile in rural Botswana (ODA 1993).

Livestock Expansion

Resource use over extensive areas of Botswana has shifted dramatically over the last century from wildlife to a livestock-dominated system. Areas in the easternmost quarter of the country, the hardveld, have changed over this period from important wildlife refuge areas to ones that today are almost totally dominated by domestic stock. It is a trend that to a large extent has been paralleled in the remaining Kalahari sandveld area of Botswana, with the availability of water in shallow depressions or pans providing the initial foothold for domestic stock to graze Kalahari pastures. Over the last forty years borehole technology has broken down this restriction by tapping deep aquifers and has enabled livestock to expand deep into the sandveld (Cooke 1985).

The history and nature of Botswana's ranching programme has been comprehensively covered (see Sandford 1983; Abel and Blaikie 1989; Hubbard 1986). Suffice it to say that colonial desires to build up the national herd, which had been decimated by the rinderpest epizootic that swept through southern Africa in 1896–7, remained largely frustrated by repeated droughts and Foot-and-Mouth Disease outbreaks. Apart from improvements in animal health, two developments were to be particularly critical in providing the means for a period of unprecedented growth in the livestock sector after Independence in 1966. These were the establishment of world beef export markets via the creation of the Botswana Meat Commission (BMC) (Morrison 1986): a parastatal with control over 90 per cent of the purchase and slaughter of the nation's cattle, and the provision of borehole water sources.

The livestock industry has been promoted by support from the then European Economic Community in the form of increasingly privileged access to European export markets since 1972. Under the Beef Protocol Agreement Botswana exports beef to Europe at well above its market value, provided stringent livestock disease control measures have been met. Most controversially the latter has focused upon the provision of veterinary cordon fences with devastating consequences for Botswana's wildlife populations (Perkins and Ringrose 1996). The cattlepost expansion that has led this trend is discussed before moving on to the current status of wildlife in Botswana, the key issues, and the way forward.

Cattleposts

Cattleposts comprise a borehole, kraals—fenced or thorn bush enclosures where the cattle are kept at night—and the huts of herders. The water-point includes the borehole itself, a storage tank, and the water trough, though the quality of this infrastructure can vary considerably, with some boreholes pumping directly into a flooded water trough area. Infrastructure costs are therefore minimal, unlike the case for ranches, where due to the remote and inaccessible nature of the Kalahari, establishing and maintaining fences, paddocks, and water reticulation systems can be considerable (Sweet 1986).

Under the cattlepost system, the 'borehole is the herder' (Jerve 1982) with their operation displaying a uniform rhythm of night kraaling, and release in the morning following milking, with the herd returning to the borehole in the late afternoon. It is a remarkably simple system, with routine herding confined to the collection and kraaling of animals around the water-point at dusk, and their subsequent release in the morning, in a daily cycle that is clearly adapted to avoid working in the extreme heat of the Kalahari days, and is more generally based upon the minimum expenditure of energy (Abel *et al.* 1987).

Changes Associated with Livestock Expansion

Changes to the surrounding rangeland following the introduction of domestic stock occur via the 'piosphere effect' (Lange 1969; Andrew 1988). With increasing distance from the borehole there is an exponential increase in the amount of grazing area available. Consequently, livestock impacts are patterned to reflect a trend of extremely high herbivore pressure at the water-point itself, to very low impacts several kilometres away. Herbivore pressure refers to the combined grazing, excretion, and trampling effects of domestic stock, which in the Kalahari have been observed to result in the following zones:

1. The sacrifice zone (0–400 m), a bare-ground zone due to trampling pressure.
2. The bush-encroached zone (200–2,000 m), an area where bush cover is typically greater than 45 per cent.
3. The grazing reserve,—an area where typical tufted perennial grasses dominate due to the low herbivore pressure.

The above radii are based upon research in the eastern Kalahari (Perkins 1991; Perkins and Thomas 1993*a, b*). Without exception the area immediately around the water-point experiences heavy trampling and grazing pressure and appears as an open expanse of bare ground. Wind-blown sand creates the impression of a devastated area, with Stoddart *et al.*'s (1975) concept of a 'sacrifice zone' appearing particularly relevant, as recovery of the vegetation seems unlikely. Beyond this zone, the dominant process is that of bush encroachment, which quite simply results in an increase in shrub cover and/or density at the expense of the grass layer.

Whether or not bush encroachment is reversible (see Ch. 5), its occurrence results in a habitat with a low aesthetic value that severely compromises any future attempts to diversify the economy through such wildlife-related ventures as eco-tourism and game viewing. Indeed, in this respect wild and domestic stock are not fully substitutable as the habitat changes coincident with permanent livestock keeping may be reversible only over a relatively long period of time (i.e. sixty years or more). Pristine wildlife habitats therefore have an aesthetic value that is quite simply being lost in areas heavily stocked with livestock.

Wildlife Declines

There is an urgent need to place the status of Botswana's wildlife populations in a modern-day context. Assessment of actual numbers is problematic (GRM 1986; FGU 1988), but the large populations reported in the late 1970s by the Country Wide Animal and Range Assessment Project (CWARAP) (DHV 1980) and widely quoted (e.g. Cooke 1985), are not relevant today (Bonifica 1992; DWNP 1994*a*, *b*). Migratory species such as blue wildebeest (*Connochaetes taurinus*) and red hartebeest (*Alcephalus buselaphus*), which formed the bulk of the large herbivore biomass in the Kalahari, have suffered as much as 90 per cent mortality (Murray 1988; Patterson 1987).

The key ungulate species in the Kalahari (Table 4.1) underwent a drastic decline in numbers in the 1980s, from which they will probably never recover (Spinage and Matlhare 1992). It is a situation that contrasts markedly with the healthy and abundant populations that dominated large herbivore biomass in the Kalahari only twenty years ago. Fences played a leading role in ensuring that the declines, that form a natural part of any drought cycle, were both steeper and more absolute than in even more severe droughts of

Table 4.1. Kalahari wildlife population numbers

Species	1978	1994	Comments
Wildebeest	315,058	17,934	Decimated—longest migration in Africa irrevocably lost. Fences, drought, loss of refugia—cattlepost expansion
Hartebeest	293,462	44,737	Decimated—recovery unlikely. Fences, drought, loss of refugia—cattlepost expansion
Gemsbok	71,423	85,368	Confined to protected areas. Hunting, habitat displacement, e.g. pans lost to livestock
Eland	18,832	11,757	Fragmented populations. Loss of habitat, hunting
Springbok	101,408	67,777	Hunting
Ostrich	92,286	27,744	Fences, hunting, live capture
Kudu	6,429	7,849	Cryptic browser
Zebra	100,295	20,863	Loss of refugia—cattlepost expansion, fences

the recent past (Perkins and Ringrose 1996). While populations of the key wild ungulate populations in the central Kalahari have increased at close to their intrinsic rates over the last decade (DWNP 1998), it is important to emphasize that this is a trend symptomatic of populations that have previously crashed precipitously.

Wildlife Distributions in the Kalahari System

The Country Wide Animal and Range Assessment Project report focused exclusively upon the Kalahari and Makgadikgadi systems. While facing substantial difficulties relating to the scale, remoteness, and time constraints, the DHV (1980) report explicitly identified a number of characteristics as critical to the future conservation of the wildlife resource:

i. . . . the Kalahari appears to be a single system in which the Schwelle running north-west–southeast through the middle of the region forms an axis, about which are centred the greatest animal numbers. (iv. 21)

ii. What is certain is that the two very large populations of hartebeest and wildebeest display one outstanding characteristic: *mobility*. It must be assumed that this nomadism is one of the preconditions of their success which means that to curtail it is to risk what might well prove massive reductions in their populations. (iv. 22)

In this respect the Kalahari is like other semi-arid ecosystems in which mobility of the ungulate biomass is the key to its survival in an ever-changing environment. Rainfall plays a primary role, with a direct correlation existing between annual rainfall and large herbivore biomass (LHB) (Coe *et al.* 1976). However, in the Kalahari, even if its low soil nutrient status is incorporated (East 1984), the expected LHB for the Central Kalahari Game Reserve ($732 \, \text{kg}^{-1} \, \text{km}^2$) is well below that actually found ($400–39 \, \text{kg}^{-1} \, \text{km}^2$, Murray 1988). The latter author attributes this low LHB in the Kalahari to two factors. The first is the extreme variability in rainfall, with large herbivore populations adapted to the utilization of low primary production, and excess production being removed by fire. The second cause of low LHB is the high proportion of the total biomass occurring in the rodentivore and insectivore communities and their prey populations, with termites (*Hodotermes sp.*) particularly significant in both their contribution to total animal biomass and removal of plant material.

The DHV (1980) report highlighted the importance of comparatively slight and stochastic events, such as isolated rainfall showers in causing dramatic, short-lived changes in animal distributions. The flushing of green shoots on recently burnt ground is another well-known example, attracting both domestic and wild herbivores in large numbers (Pratt 1967), and verified by both the DWNP (Verlinden, pers. comm.) and CWARAP (DHV 1980) surveys. The latter point to the green flush associated with an isolated rainfall event in July 1979 in the southern Kalahari region which led to the concentration of some 40,000 wildebeest (*Connochaetes taurinus*) and 25,000 eland (*Taurotragus oryx*). By contrast the three previous counts over a broader southern Kalahari area had returned an average of 4,500 wildebeest and 6,000 eland (from DHV 1980: iv). This nomadism clearly has an important bearing for the strategy of both wildlife and livestock production on Kalahari pastures, which are discussed further below.

Wildlife Population and Movement Patterns

Wildlife species that are dependent upon surface water are either confined to northern Botswana (e.g. buffaloes) or able to move between this region and the Makgadikgadi Pans Game Reserve (e.g. elephant and zebra) as part of a seasonal migration. In this respect, the Kalahari wildebeest is exceptional in that it is able to meet its water requirements from plants, especially tsamma melons (*Citrillus lanatus*) and the gemsbok cucumber (*Acanthosicyos naudiniana*), in the dry season (Murray 1988; Lindsay 1992). The latter points to the existence of a 'migrant' population and also a smaller resident subpopulation (a few hundred animals), centred on the Central Kalahari Game Reserve, with more restricted movements.

Like many Kalahari species, such as gemsbok and springbok, wildebeest are dependent upon pans and fossil river valleys for the essential minerals that their soils provide (Parris and Child 1973). As such the so-called 'Schwelle' region is a key resource area for the wildebeest. Its concentration of mineral pans, together with open woodlands, constitutes a 'core area' in the wet season, when calving takes place, with the wildebeest dispersing, often along dry river valleys, as the dry season progresses (Williamson and Williamson 1985*a*, *b*; Lindsay 1992).

However, in drier than average years the wildebeest must seek access to water, which means either moving southwards to the Orange/Molopo River, or north-eastwards to Lake Xau, from the southern and western Kalahari. Both sets of migrants have met with catastrophe during the movements in droughts that have been documented in the available historical records (Lindsay 1992; Perkins and Ringrose 1996). The Kalahari wildebeest does not fit well into a classical migratory, or water-dependent, classification that is bestowed upon eastern African populations. Unlike the wildebeest population in the Serengeti of Tanzania there is no annual trek along fixed routes, although in terms of distance covered the range of the population in the central and southern Kalahari is comparatively much greater (Lindsay 1992).

The first comprehensive study of large herbivores within the Kalahari ecosystem (DHV 1980) concluded that the wildebeest population, estimated at over a quarter of a million, were essentially nomadic and opportunistic in their spatial and temporal movements, according to the availability of green grass and melon crops (Lindsay 1992). It seems that in average to good years, moisture from such foods enables them to reside in the Schwelle region throughout the year, while in drier years they move outwards to the north and north-east and also to the south in search of surface water (ibid.). In either case, such migrants met with disaster in the severe 1980s drought, most notably along the Kuke fence.

Today water provision, from ten pumped boreholes concentrated within the pan and valley systems of the northern and southern CKGR (see also Verlinden 1994), is seen as a solution to the migrant wildebeest loss of access to water from the Boteti River. However, artificial water provision within the Kalahari ecosystem, while appearing to be the logical solution, creates a number of fundamental conservation dilemmas, not least the disruption of a wilderness area (Sweet 1986; Lindsay 1992). The fundamental ecological issues involved, include the disruption of resident Kalahari water-independent herbivores through grazing depletion, the attraction of predators, and significantly the increased possibility of disease transmission (e.g. anthrax) (Lindsay 1992). Indeed, the benefits of maintaining the Kalahari as a waterless environment may well outweigh the ecological risks and financial costs involved in fully implementing the borehole strategy.

The plight of the Kalahari wildebeest illustrates the more general trend of wild ungulates becoming increasingly confined to the protected areas, through hunting and livestock-related displacement. More intensive management (borehole provision and fencing) is effectively seen as the panacea to stabilizing and conserving the wild animal biomass in the protected areas of the Kalahari, as livestock continues to expand around them.

Regrettably such developments ignore the inherent variability of semi-arid ecosystems and in particular the occurrence of pronounced and severe droughts. Policies in both the livestock and wildlife sectors operate at the wrong spatial scale and ignore the value of multi-species projects. The latter were meant to characterize developments in the buffer zones, or Wildlife Management Areas (WMAs), that make up 22 per cent of the country, and adjoin Game Reserves and National Parks, but have generally been slow to develop.

Wildlife Management Areas

Wildlife Management Areas are meant to provide livelihood opportunities based on traditional practices and act as conservation areas (see Ch. 7). Multifarious reasons may be cited for the limited success of the WMA concept in the Kalahari, not least the lack of political will to gazette WMAs and to implement conservation policies in a meaningful fashion. The reality has been encroachment of livestock into them and conflict between the conservation of the wildlife and veld food resources on the one hand, and the provision of meat and milk from domestic stock on the other. Again it should be emphasized that this scenario was foreseen well in advance, with DHV (1980: 44) pointing out that,

If the WMA concept is not used immediately as the framework within which to insti-tute extensive game management, then the whole principle will rapidly collapse and the Kalahari continue to be overrun with cattle and goats.

and further,

the integrity of the Kalahari as a major wild animal system is at high risk. The key to its integrity is the Schwelle. The greater part of the game herd—the hartebeest, wildebeest, ostrich, springbok and kudu—require both access to the Schwelle and freedom of passage across it. Likewise it will be the site of much of the livestock expansion of the immediate future.   (p. 38)

Much of the range in the Schwelle is contained within the Kgalagadi Wildlife Management Areas. The need to give conservation a high priority within this region has been stated repeatedly (Botswana Society 1971; DHV 1980; Murray 1988; Cumming and Taylor 1989; MLGL 1991; RPM 1995). The message has not changed and today the predictions embodied in these works are being borne out:

The WMAs are coming under increasing pressure from cattle interests because of crowding in the Communal Areas and inactivity in the WMAs. Most cattle grazing is along the Ghanzi-Kang road, where people have illegally taken up Roads Department boreholes for their own private use, some of which were mistakenly allo-cated to cattle-owners by the Ghanzi District Council. The cattle can travel up to 40 kms away from the cattleposts. Further grazing pressure has resulted with the alloca-tion of boreholes which are inconsistent to current Land Board policy for allocation within these areas.   (RPM 1995: i. 26).

The fate of WMAs has always rested upon a political decision, or non-decision. The latter has ensured that the encroachment of borehole-led cattleposts has remained unchallenged, even though the boreholes have either been illegally drilled by wealthy individuals and/or mistakenly allocated by Land Boards. Undeniably, the benefits of such actions have accrued to a few individuals rather than the local communities of the area (MLGL 1991). The completion of the Trans-Kalahari Road between Ghanzi and Gaborone hås intensified the conflict between livestock and wildlife, and demands to fence the road, if met, will result in the severance of the central Kalahari ecosys-tem from that of the south-western system. Like other fencing developments this would have major consequences for Kalahari wildlife.

The Impact of Fences

Taylor and Martin (1987) noted that any less-developed country that aspires to export beef to international markets is required to meet high standards of veterinary hygiene and disease management. Particularly in response to the generous tariff concession exercised by the EU, the Botswana government endorsed its commitment to exporting beef to European markets by implementing stringent veterinary measures. Undeniably these measures, and in particular the fencing component, have conflicted with other land uses especially those relating to wildlife conservation and utilization.

Despite the recent outbreak of lung disease (CBPP) in north-western Botswana in 1996 that resulted in the destruction of some 350,000 cattle, it is Foot-and-Mouth Disease (FMD) that is potentially the most damaging of all diseases to the livestock sector (Taylor and Martin 1987). The latter authors point out that in contrast to temperate countries FMD has limited pathogenic effect in tropical environments. It is primarily a disease of economic importance because its outbreak terminates beef exports.

Foot-and-Mouth Disease is endemic in Zimbabwe, especially in the area adjacent to Gonarezhou National Park (ibid.), with the few outbreaks that have occurred in Botswana explicitly regarded as having been transmitted by the illegal movement of cattle across this border or the Caprivi Strip of Namibia (McGowan International and Coopers Lybrand 1988). Buffalo (*Syncerus caffer*) are the most important host of FMD, with the buffalo areas of northern Botswana and bordering regions with Zimbabwe regarded as a permanent reservoir of the disease and so to be high-risk zones. Consequently, the disease control programmes that have been applied have focused upon the total exclusion and/or eradication of buffalo from cattle-producing and non-protected areas in Botswana (and Zimbabwe). Although a vaccine has been developed against FMD and is administered bi- or tri-annually in the high-risk zones, beef importers will not accept vaccination as an effective control, and insist upon fences as the only reliable means of containment (Taylor and Martin 1987).

In Botswana this has led to the development of an extensive network of veterinary cordon fences throughout the country, established in the interest of national policy and consistently with little advance planning and negligible liaison with ecologists and land-use planners. The country is effectively divided into seventeen zones with double or single cordon fences as boundaries (Fig. 4.1). These, together with general movement restrictions and quarantine camps along them, control the intra-zonal movement of livestock and other animals. However, such an obsession with the retention of the EU market continues to incur very high costs in capital and recurrent expenditure (McGowan International and Coopers Lybrand 1988) and has had severe ramifications for migratory ungulates. As indicated above, these declined precipitously after the fences were erected. In a conclusion that applies with striking relevance to Botswana, it has been pointed out that:

Under pressure from beef importers, it would be all too easy to adopt a policy that attempted to make the entire country disease free for cattle production, regardless of

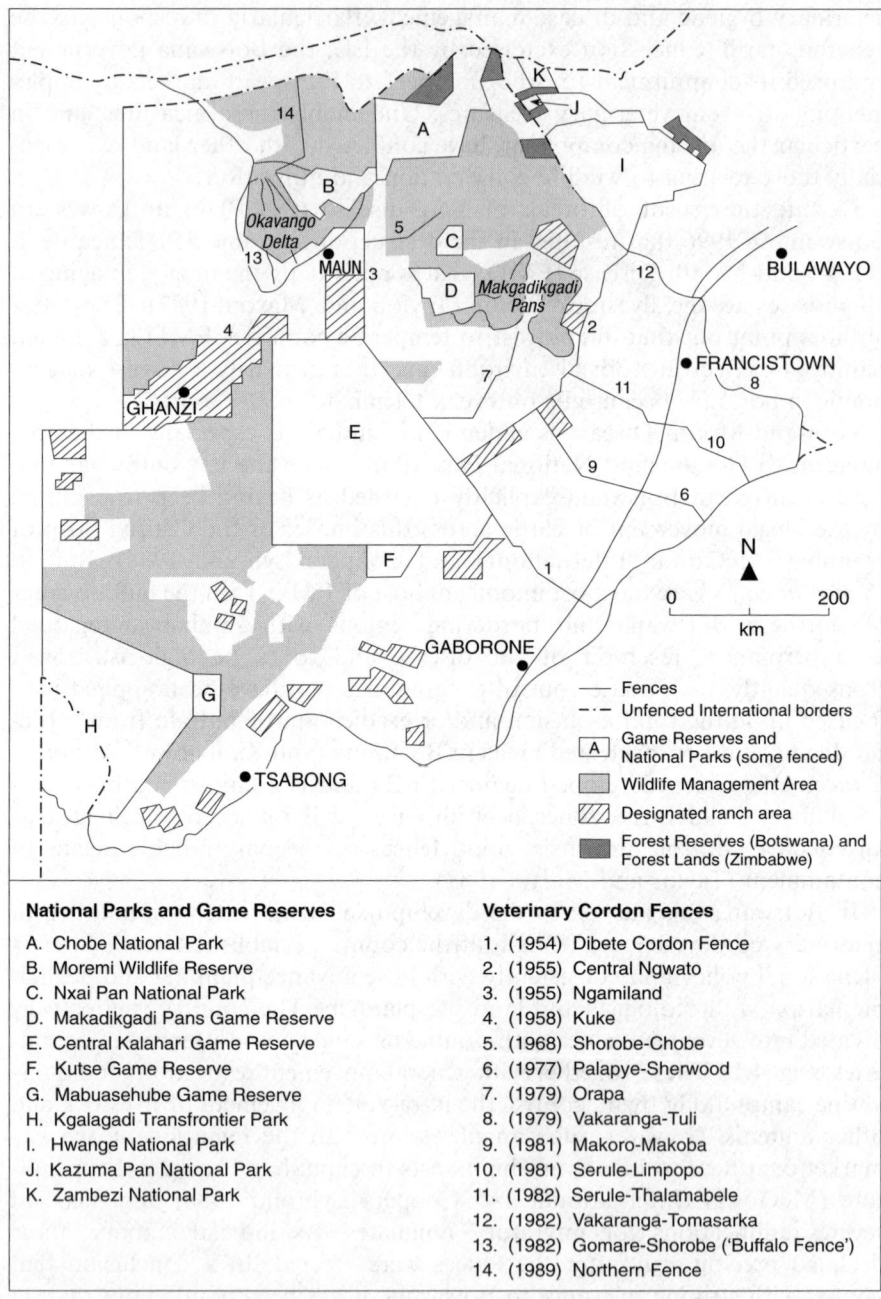

National Parks and Game Reserves

A. Chobe National Park
B. Moremi Wildlife Reserve
C. Nxai Pan National Park
D. Makadikgadi Pans Game Reserve
E. Central Kalahari Game Reserve
F. Kutse Game Reserve
G. Mabuasehube Game Reserve
H. Kgalagadi Transfrontier Park
I. Hwange National Park
J. Kazuma Pan National Park
K. Zambezi National Park

Veterinary Cordon Fences

1. (1954) Dibete Cordon Fence
2. (1955) Central Ngwato
3. (1955) Ngamiland
4. (1958) Kuke
5. (1968) Shorobe-Chobe
6. (1977) Palapye-Sherwood
7. (1979) Orapa
8. (1979) Vakaranga-Tuli
9. (1981) Makoro-Makoba
10. (1981) Serule-Limpopo
11. (1982) Serule-Thalamabele
12. (1982) Vakaranga-Tomasarka
13. (1982) Gomare-Shorobe ('Buffalo Fence')
14. (1989) Northern Fence

Fig. 4.1. Veterinary fences in Botswana and their interaction with principal wildlife migration routes.

land capability. Wildlife populations would be reduced to those in a few heavily cor-doned enclosures, grudgingly accepted to satisfy the conservation lobby. There is little reason for the developed world to influence land use in the marginal areas [of Zimbabwe] especially if it is short-term economic exploitation at the expense of land. (Taylor and Martin 1987: 333)

The Veld Food Resource in the Kalahari

Meat and mongongo for strength, honey for sweetness, and wild orange for refreshment.

(Ju/hoansi reply to questions about
an the ideal diet, quoted from Lee 1984: 48)

A large portion of the literature relevant to the utilization of veld foods in the Kalahari comes from studies relating to the lifestyle and gathering activities of the Khoisan people. Indeed, the veld food plants used by hunter-gatherer communities have formed part of an extensive and well-documented botanical and anthropological research, particularly in the Central Kalahari Game Reserve of Botswana (Tanaka 1976; Silberbauer 1981) and the Nyae Nyae region of Namibia (Story 1958, 1964; Lee and DeVore 1976; Lee 1979, 1984).

The early studies (e.g. Story 1958) provide an impressive inventory and description of some of the plants documented on pioneering botanical surveys, while the works of anthropologists (for example Tanaka 1976; Lee 1979, 1984) generally place the emphasis upon ethnobotany. All point to the remarkable depth of ecological knowledge possessed by hunter-gatherer communities, as reflected in the number and diversity of plants used, and the ability to locate them from seemingly imperceptible signs.

Naturally all the anthropological studies have exclusively focused upon the subsistence use of veld foods, and have identified complex criteria that are used to arrange plant foods into a hierarchy of classes of desirability. Lee (1984), for example, points out that over 100 species of wild plants are clas-sified by the Ju/hoansi as edible. Factors such as their abundance, the dura-tion of the eating season, ease of collecting, nutritional value, and tastiness are all important in determining whether they are 'strong' or 'weak' foods. Using these factors, together with data on availability, Lee (1984) recognized a six-class hierarchy of foods (Table 4.2), in a classification that shows close parallels with that developed by Tanaka (1976) for the G/wi and G//ana of the Central Kalahari Game Reserve of Botswana.

Many veld products are valued for their ability to provide water, and to a lesser extent food, in the dry season on a subsistence basis. Consequently, since the advent of borehole technology has overcome the age-old problem

Table 4.2. Hierarchy of foods

Food Class	No. of Species	Criteria
6. Primary	1	Widely abundant year round
5. Major	13	Widely abundant
4. Minor	19	Locally seasonally abundant
3. Supplementary	30	Locally seasonally available
2. Rare	19	Rarely observed to be eaten
1. Problematic	23	!Kung classify as edible; not observed to be eaten

Source: Adapted from Lee (1984: 47).

of a lack of surface water in the Kalahari, the significance of such water-bearing veld foods has declined for many former hunter-gatherer groups. Furthermore, the wage-earning opportunities that have followed borehole provision (for example, cattle herding) have also ensured that modern food-stuffs have increasingly substituted even some of the more important veld foods. From a list of 333 species Arnold *et al.* (1985) identified sixty-six species that showed some potential for economic exploitation on the basis of their nutrient composition, and then narrowed this down further to twenty-seven species. Only a summary (Table 4.3) is included here, with Arnold *et al.* (1985) also incorporating Lee's (1984) ranking (Table 4.2) and dividing them into seven life-form categories (A–G). The resources in Table 4.3 contribute to rural livelihoods in a number of ways. They bolster food security, contribute to drought coping mechanisms, generate income, and they are used by rural poor and female-headed households (LaFranchi 1996). These points are closely interrelated and often overlooked, or grossly underestimated in conventional economic studies, as they are all difficult uses to fully account for (ibid.). Moreover, consideration of the commercial value of veld products usually, focuses attention on income generation, whose potential over much of the Kalahari must be stated to be limited (Perkins 1999). However, the other four points all play a vital role in poverty alleviation in rural areas and undoubtedly decline as cattle-keeping moves into areas on a permanent basis.

Impact of Livestock upon Veld Foods

In the Kalahari, resource areas, or *n!oresi*, typically centre on low-lying depressions or pans for their water-providing potential. Pans play an important role within the Kalahari ecosystem and are of vital significance to wildlife. Their relatively high silt-clay fractions offer both important minerals, and hence otherwise unobtainable soluble salts (Parris and Child 1973), and also the possibility of water, either standing following rain, or attainable through the digging of shallow wells. For the same reason such habitats are typically used to site boreholes and so become the focal point of livestock impacts and conflict with the hunting and gathering resources.

The wild plants currently known to be used for subsistence and medicinal purposes in Botswana include a number of taxa across a diverse array of plant parts, that includes tubers, roots, fungi, bark, gums, resins, stems, leaves, fruits, flowers, and seeds. The medicinal value of many veld products has yet to be formally discovered, and there is an absolute dearth of information concerning the impact of livestock upon veld foods. Early inventories of veld food resources (Taylor and Moss 1983) and consideration of the effects of livestock expansion upon rural economies (Campbell 1981), all point to the depletion of the veld food resource, although hard data is lacking.

Kgabung's (1999), work in the Kgalagadi WMAs represents one of the earliest attempts to test this hypothesis, by assessing how the nature of the veld food resources changes outwards from boreholes used for livestock-keeping (Table 4.4). Significantly, the wealth of medicinal plants that are utilized by local communities were not considered, simply because of the great diversity of plant parts utilized, and the difficulties in attaching scientific names to them. Nevertheless the findings pointed to the loss and/or depletion of the key veld food resources with increasing stocking rates around the three boreholes studied.

In summary, the abundance and diversity of most wild foods and thatching grasses increases with distance from the water-point. While the grazing gradient was clear at two of the boreholes where stocking rates were relatively high, the final piosphere studied was grazed only by smallstock and donkeys. In this case, the high amounts of standing grass biomass had constituted a fuel load with a severe fire impacting upon the veld food resources at the time of the study. Indeed, this result serves to emphasize the interactive effects between grazing pressure and fire occurrences and the complexity of the veld food resource availability issue.

The Changing Use of Fire

It can be emphasized that traditional ecological knowledge has long formed the foundation for the use of the Kalahari rangelands for subsistence hunting and foraging. Indeed, it is clear from the wealth of Khoisan literature available that fire has been a basic tool in the management of Kalahari rangelands by hunter-gatherer populations for at least the last forty thousand years. Despite this the changing role of fire in the Kalahari ecosystem appears to have been largely overlooked in the literature. By contrast, the importance of the Aboriginal use of fire in shaping Australian ecosystems is a common theme. Flannery (1994: 217) for example quotes Giles (1889), in emphasizing

Table 4.3. Species showing potential for economic exploitation

Plant name	Life form	Plant part	Potential nutrition[a]	Domestication potential[b]	Yield estimate[c]	Desirability estimate[d]	Total	Utilization[e]	San rating
A									
Ricinodendron rautanenii	Tree	Fruit	5	2	6	6	360	W (D)	5
Sclerocarya birrea	Tree	Fruit	2	3	6	6	216	W (D)	5
Hyphaene ventricosa	Tree	Fruit	5	3	5	2	150	W (D)	(3) 5
Strychnos cocculoides	Tree	Fruit	1	3	4	5	60	W (D)	4
Grewia retinervis	Shrub	Fruit	3	3	2	3	54	W	(3) 5
Adansonia digitata	Tree	Fruit	5	1	3	3	45	W	(4) 5
B									
Cucumis metuliferus	Herb	Fruit	2	6	5	3	180	D	—
Cucumis africanus	Herb	Fruit	2	5	4	3	120	D	—
Cucumis anguria	Herb	Fruit	(1)	6	5	4	120	D	4
Pentarrhinum insipidum	Herb	Fruit	2	5	3	3	90	D	4
Acanthosicyos naudiniana	Herb	Fruit	2	3	5	3	90	D	(3) 5
Citrillus lanatus	Herb	Fruit	1	6	5	3	90	D	(4) 5
Coccinia sessilifolia	Herb	Fruit	1	5	4	3	60	D	3
Coccinia kalahariensis	Herb	Fruit	1	4	4	3	48	D	5
Coccinia rehmanni	Herb	Fruit	1	4	4	3	48	D	(4) 5
Acanthosicyos horrida	Shrub	Fruit	1	2	3	3	18	W (D)	—
C									
Ricinodendron rautanenii	Tree	Nut	6	3	6	6	648	W (D)	5
Sclerocarya birrea	Tree	Nut	6	2	6	6	432	W	5
Guibortia coleosperma	Tree	Nut	5	3	4	5	300	W	(2) 5
Adansonia digitata	Tree	Nut	6	1	3	3	54	W	(4) 5

D								
Tylosema esculentum	Shrub	Nut	6	5	5	900	D	5
Citrillus lanatus	Herb	Seed	6	5	5	540	D	(4) 5
Bauhinia petersiana	Shrub	Seed	4	5	4	288	D	(3) 5
Schotia afra	Shrub	Seed	3	4	3	216	D	—
Acanthosicyos horrida	Shrub	Seed	2	3	6	180	W (D)	—
E								
Trachyandra falcata	Herb	Infl.	3	4	4	192	D	—
F								
Pentarrhinum insipidum	Herb	Leaves	4	5	2	160	D	4
Cucumis africanus	Herb	Leaves	3	5	2	120	D	—
G								
Pelargonium incrassatum	Herb	Tuber	3	3	4	144	D	—
Tylosema esculentum	Shrub	Tuber	1	6	4	120	D	5
Coccinia sessilifolia	Herb	Tuber	1	6	4	120	D	3
Coccinia rehmannii	Herb	Tuber	2	4	3	96	D	(4) 5
Babiana dregei	Herb	Corm	3	3	3	81	D	—
Vigna lobatifolia	Herb	Tuber	1	5	4	80	D	4
Cucumis kalahariensis	Herb	Tuber	5	5	4	80	D	5
Fockea angustifolia	Herb	Tuber	1	4	4	64	D	5
Cyanella hyacinthoides	Herb	Corm	2	3	3	36	D	—

Source: Arnold *et al.* (1985: 74).

[a] Nutrient Composition: based on the number of chemical constituents which have a value greater than 20% of the daily requirement for that constituent. Scoring was as follows: 0–1 constituent = 1, 2–3 = 2, 4–5 = 3, 6–7 = 4, 8–9 = 5, 10–12 = 6.

[b] Domestication potential: this collectively takes into account: time to first crop, ease of harvesting, accessibility of edible part, distribution (width of habitat), ease of propagation, handling and storage qualities.

[c] Relative yield: this takes into account two aspects: the size of the edible part and the number of edible parts per individual plant.

[d] Desirability: this criterion reflects the desirability or potential acceptability of the plant part as a food and mainly takes into account palatability and the nature of the edible part.

[e] W = utilization in the wild and D = domestication.

Table 4.4. Changes in the nature of veld food resources outwards from live-stock boreholes in Kgalagadi district

	Species Composition	
	Community 1 (closer to the water-point)	Community 2 (some distance from the water-point)
Veld product (wild foods and thatching grasses)	Mainly: Ruderals (e.g. *Aristida meridionalis*, Mosumo, Tshoba, *Pagodus* spp., etc.) Pan species (e.g. *Kedrostis hirtella*, etc.) Other (e.g. *Ziziphus mucronata* etc.)	A variety of wild foods (e.g. *Grewia retinervis, Grewia flava, Citrullus naudinianus, Coccinia rehmannii*, Sekhwande, etc.) Thatching grasses include *Stipagrostis uniplumis* and *Eragrostis uniplumis*
Herbaceous plants	*Cynodon dactylon, Sporobulus* spp., *Senna italica Tribulus terrestris*, sunflower-like weed, *Aristida congesta*, and forbs	Herbs include forbs and species such as *Stipagrostis uniplumis,* *Urochloa trichophus*, Seloba, Ngwanyane, etc.
Shrubs or bushes	*Acacia mellifera, Gnidia polycepala, Acacia hebeclada*, and pan species such as *Cactophrates alexandre, Monechma spp.*, etc.	Examples of shrubs are *Grewia flava, Grewia retinervis, Ziziphus mucronata, Acacia luderitzii, Acacia erioloba*, and *Rhus ternuinervis*
Trees	*Ziziphus mucronata, Acacia mellifera, Acacia fleckii*	*Terminelia sericea, Acacia spp., Boscia albitrunca*, etc.

the inseparability of fire and Aborigines: 'The natives were about, burning, burning, ever burning; one would think that they were of the fabled sala-mander race, and lived on fire instead of water.'

The deliberate use of fire by Aboriginals to maintain optimum rangeland habitats for wildlife and invertebrates and also to safeguard against uncon-trolled, destructive wildfires has been frequently stated (Young and Ross 1992). Even so the environmental impact of land alienation with the arrival of the Europeans, and the banning of Aboriginal 'fire-stick farming' because of the general perception that fire was dangerous and destructive to the graz-ing resource, is sometimes overlooked (Flannery 1994).

With the recognition of Aboriginal land rights in some areas has come the realization that the Aboriginal use of fire has played an important part in the conservation and management of national parks in central Australia, by both regenerating the natural vegetation and creating suitable habitats for wildlife species (Young and Ross 1992). Indeed, the latter authors point to the return of Aborigines to their homelands in the Tanami Desert, after a thirty-year absence, and the reintroduction of traditional mosaic burning practices to restore the land from the 'rubbish state' in which they found it. Striking parallels should perhaps be drawn with the San's use of fire in the Kalahari rangelands of southern Africa, which seem likely to have undergone a similar radical change in burning regimes following official bans on the use of fire, and the drastic wildlife declines experienced in the early 1980s. This is not to suggest that fire events are no longer a factor in the Kalahari as the opposite is true. Instead, it seems likely that recent fires in the Kalahari (i.e. since the 1980s drought) are more severe and widespread than they were in the 1960s and 1970s, because of the coincidence of two factors. First, the current, almost unprecedented low in large wild herbivore biomass in the Kalahari system, perhaps surpassed only by the rinderpest epidemic one hundred years ago, has meant that there is a large, dead biomass of standing grass after good rainfall years on wildlife-dominated rangelands. Second, the ban on veld fires has resulted in the general absence of hunter-gatherer burning practices, which seem likely to have created a greater diversity of spatial and temporal burn mosaics, than is now the case.

The use of fire by hunter-gatherer communities in the Kalahari is also borne out by numerous anthropological studies. In this respect it is perhaps ironic that no data on fire is currently forthcoming from the communities, despite the fact that it is one of the most important ecological determinants of the state and dynamics of semi-arid savanna vegetation. Indeed, the fact that burning practices have changed dramatically in recent decades following the government's ban on burning, coincident with a shift from hunting and gathering to livestock-dominated modes of production, appears to have been largely overlooked in Botswana. This is unfortunate, as is the general suspicion with which the communities meet any discussion of fire ecology.

In Nyae Nyae, Powell (1994), reporting on an Environmental Planning Committee meeting in Eastern Bushmanland, neatly captures the polarization of views on fires and fire management, in quoting the Department of Forestry's representative:

the Ministry sees a problem with burning every year as it stops trees regenerating. We in the Ministry feel that fires should be managed. I am not suggesting that fires should be stopped altogether, but every time a member of the community needs to burn he should ask a forestry official to come out and assist. We need to provide the community with training in how to manage fires . . . (Powell 1994: 62)

and also the community's view,

Bushfire is really important to us because that's what we are living from; and it's not just us, it is also other things e.g., vegetation to look good, grazing, hunting, bush food and many others. Anyone who is thinking of stopping fire, because that's our method of managing our veld will have to take over our management role. When we knew our grandparents we were taught how to manage the land with fire. Anyone who thinks of stopping fire must be prepared to help our children and our children's children. (ibid. 63)

Such exchanges could well have occurred in Botswana, where there is an equal need to discuss the use of fire openly and without fear of litigation. Fire is clearly an essential tool for the management of Kalahari rangelands and yet one that appears to have been grossly neglected in our attempts to understand how livestock expansion has changed the dynamics of the Kalahari ecosystem.

The Mechanism for Change in the Kalahari

The next decade will prove a critical period for the Kalahari of Botswana and the wildlife populations that it supports. The potential for multi-species projects is undoubtedly declining as borehole-based livestock production continues to expand into the Kalahari. This movement appears to have been accompanied by a pronounced and interrelated trend in critical environmental and socio-economic parameters. We believe that the main pattern is:

1. A borehole is drilled and livestock are grazed around it.

2. Output per hectare increases initially as food is available from a diverse array of sources. Veld products around the borehole can be gathered, wildlife hunted, and milk (and meat—mainly from goats) is available, particularly in good rainfall years. Output from veld products may actually be above that found on rangeland ungrazed by domestic stock, due to the beneficial effects of disturbance on the production of some important veld foods.

3. Stocking rates increase and livestock impact upon the range via the piosphere effect such that distinct zones of impact open up, particularly a 'sacrifice zone' (0–400 m) and a bush-encroached zone (400—2000 m).

4. Wildlife populations are largely displaced from the area due to the effects of hunting and an important source of animal protein made inaccessible to communities found within the area.

5. Trampling and grazing effects by cattle around the borehole cause major changes in the structure, species composition, and cover of the surrounding vegetation. Veld foods either disappear entirely, or, if present, fail to produce a useful crop due to disturbance effects.

6. Human dependency upon water from the borehole and milk from the kraal increases, as wildlife and veld food outputs decline drastically.

7. Livestock ownership becomes concentrated into the hands of borehole owners, often syndicates, who benefit from livestock sales, and the ready availability of labour, in the form of former hunter-gatherers and Kalahari residents.

8. Fencing of the rangeland around the borehole becomes necessary as borehole densities increase and livestock owners seek to gain exclusive use of the surrounding pastures.

9. Kalahari residents become entirely dependent upon 'drought' and/or 'destitute' relief, following the loss of wildlife and veld food resources.

10. An open and flexible system characterized by mobility and multiple use has become a closed cycle characterized by a single dominant land use (livestock) and structural poverty.

We believe that this has been the overwhelming sequence of events in much of the Kalahari of southern Africa, and more specifically upon Kalahari pastures in Botswana. Drought and fencing (particularly for disease control) have all accentuated this trend, as has the effective EU subsidy, via the Beef Protocol Agreement. Income disparities have therefore been accentuated with livestock-related wealth concentrated into remarkably few wealthy households. Communal farmers have suffered a double blow in the sense that the wildlife and veld food declines have removed important sources of subsistence food, while communal pastures have been depleted by livestock grazing from neighbouring cattleposts. The latter is known as 'dual grazing rights', and remains a major grievance amongst communal farmers, as wealthy individuals are able to exploit the communal resource before retreating to their own 'exclusive' pastures around their borehole. As a result communal land always appears overgrazed in relation to surrounding fenced private farms and has consequently provided considerable impetus to fence off communal land.

At cattle posts where people either freely admitted hunting, or where animal remains around the camps revealed a heavy reliance on hunting, there were an average of 137 cattle per borehole. Where people did not hunt but admitted gathering, there were an average of 188 cattle per borehole. Where people told us (and where visual observation supported this) that they neither hunted nor gathered, but subsisted totally on domesticated animals and plants, there were an average of 241 cattle per borehole. It is clear that, as more and more cattle are maintained in an area, as overgrazing progresses and as more people are required to help manage the beasts, Basarwa slowly stop relying on hunting and then later on gathering of wild foods for a livelihood. (Ebert *et al.* 1976, quoted in Hitchcock 1977: 268)

The pressure for continued livestock expansion in the Kalahari of Botswana has already been emphasized and has in fact only being stalled by the lack of suitable groundwater resources. Although livestock-rearing can play a valuable role in satisfying growing nutritional needs and so contribute to food security, particularly at a time when the key wild ungulates are at low densities, there is a real need to address the optimal balance between domestic and wild ungulate populations.

The vital role of domestic stock has been recognized at the Nyae Nyae Conservancy in north-eastern Namibia, but with very conservative stocking rates recommended (56ha/LSU) (Stuart-Hill and Perkins 1997). Even so, the critical stocking rate thresholds that impact upon the veld product resource are simply not known, although it is clear from recent research around the Matsheng cattleposts (Kgabung 1999) that the veld food resource cannot coexist with the stocking rates that are typically found around Kalahari cattleposts. As the DHV (1980), survey states,

New range will be pioneered by traditional cattle-posts. . . . Game will be greatly reduced and displaced into final refuges defined by the one obstacle the stockman cannot yet overcome: lack of suitable groundwater. Such a process cannot be condemned. . . . [provided decision makers are] . . . aware of the consequences. In the context of game management there is no case for planned game use if this is only opportunistic pending its substitution by livestock.   (DHV 1980: 38–9)

Moreover,

Enhanced game use is seen as the best way to raise the standard of living of the greatest number of people in the Kalahari, particularly those who are the poorest. (ibid. 45)

Conclusion

There should be little doubt that continued livestock expansion in critical wildlife areas, such as the Schwelle, will have an impact on the remaining Kalahari wildlife populations that is disproportionate to that which has occurred in other less critical habitats in the Kalahari ecosystem. Access to and preservation of mobility across the Schwelle for the key wild ungulate species, which requires connectivity between the central and southern Kalahari ecosystems, appears to be an essential prerequisite for sustained recovery of their populations. Indeed, for the communities in these areas, which constitute some of the poorest people in the country, multi-species projects across a wide diversity of activities (that includes subsistence and trophy hunting and cultural tourism) offers the only real chance of sustained poverty alleviation. Regrettably such initiatives have been slow to develop with livestock expansion skewing the income-generating potential of these areas towards a few wealthy individuals in a seemingly inexorable cycle of habitat loss and veld product and wildlife depletion.

The Kalahari ecosystem in Botswana is therefore perhaps the most threatened in the country, with the existing policies directing the conservation and utilization of its resources along an unsustainable path at the wrong spatial scale. Macro-economic assistance from the EU in the form of the Beef Protocol Agreement has played an important role in driving policy in the livestock sector. It remains to be seen whether this trend can be broken this century and replaced by a more enlightened phase of integrated ecological and socio-economic development that strikes a sustainable balance between livestock- and wildlife-based economies.

References

ABEL, N. O. J., and BLAIKIE, P. M. (1989), 'Land degradation, stocking rates and conservation policies in the communal rangelands of Botswana and Zimbabwe', *Land Degradation and Rehabilitation*, 1: 101–23.

ABEL, N. O. J., FLINT, M. E. S., HUNTER, N. D., CHANDLER, D., and MAKA, G. (1987), *Cattle-Keeping Ecological Change and Communal Management in Ngwaketse* (Norwich: ILCA/IFPP/ODG).

ANDREW, M. H. (1988), 'Grazing impact in relation to livestock watering points', *Trends in Ecology and Evolution*, 3/12: 336–9.

ARNOLD, T. H., WELLS, M. J., and WEHMEYER, A. S. (1985), 'Khoisan food plants: Taxa with potential for future economic exploitation', in G. E. Wickens, J. R. Goodin, and D. V. Field (eds.), *Plants for Arid Lands*. Proceedings of the Kew International Conference on Economic Plants for Arid Lands, Royal Botanic Gardens, Kew, England, 23–7 July 1984 (London: George Allen & Unwin), 69–86.

BONIFICA (1992), *Aerial Surveys of Botswana 1989–1991: Final Report* (Gaborone: Technical Assistance to the Department of Wildlife and National Parks).

BOTSWANA SOCIETY (1971), *Proceedings of the Conference on Sustained Production from Semi-Arid Areas with Particular Reference to Botswana*, Special Edition No. 1. (Gaborone: Botswana Society).

COE, M., CUMMING, D. H. M., and PHILLIPSON, J. (1976), 'Biomass and production of large Arfrican herbivores in relation to rainfall and primary production', *Oecologia*, 22: 341–54.

COOKE, H. J. (1985), 'The Kalahari today: a case of conflict over resource use', *Geographical Journal*, 151: 75–85.

CUMMING, D. H. M., and TAYLOR, R. D. (1989), *Identification of Wildlife Utilisation Projects for DWNP* (Gaborone: Government of Botswana).

DHV (1980), *Countrywide Animal and Range Assessment Project*, 7 vols. (Gaborone: European Development Fund and Ministry of Commerce and Industry).

DWNP (1994*a*), *Aerial Census of Animals in Botswana: Dry Season 1994*. (Gaborone: ULG Consultants Ltd. UK; Technical Assistance to the Department of Wildlife and National Parks).

——(1994*b*), *Aerial Census of Animals in Botswana: Wet Season 1994*. (Gaborone: ULG Consultants Ltd. UK; Technical Assistance to the Department of Wildlife and National Parks).

——(1998), Draft Management Plan for the Central Kalahari and Khutse Game Reserve (Gaborone: Department of Wildlife and National Parks).

EAST, R. (1984), 'Rainfall, soil nutrient status and biomass of large African savanna mammals', *African Journal of Ecology*, 22/4: 245–70.

FGU (1988), *Review of the Aerial Monitoring Programme of the Department of Wildlife and National Parks, Botswana*, Special Report of FGU-Kronberg Consulting and Engineering GMBH (Gaborone: Department of Wildlife and National Parks).

FLANNERY, T. (1994), *The Future Eaters* (Melbourne: Reed International Books).

GRM (1986), *Re-establishment of a National Range Resource Assessment and Monitoring Programme*, Draft Final Report, GRM International Pty. Ltd., (Gaborone: Ministry of Agriculture).

HITCHCOCK, R. K. (1978), *Kalahari Cattleposts: A Regional Study of Hunter-*

Gatherers, Pastoralists and Agriculturalists in the Western Sandveld Region, Central District, Botswana (Gaborone: Government Printer).

HUBBARD, M. (1986), *Agricultural Exports and Economic Growth. A Study of Botswana's Beef Industry* (London: KPI Ltd.).

JERVE, A. M. (1982), 'Cattle and inequality: A study in rural differentiation from southern Kgalagadi in Botswana', DERAP Publications, 143 (Bergen: Chr. Michelson Institute).

KGABUNG, B. (1999), 'The Impact of the Expansion of Livestock on the Distribution and Utilization of Wildlife and Veld Products in Northern Kgalagadi Sub-District, Botswana', MSc Thesis, Department of Environmental Science, University of Botswana.

LAFRANCHI, C. (1996), Small Scale Subsistence Use of Natural Resources in Namibian Communal Areas: Estimating the value to livelihood of selected wild foods, medicinals, building and craft materials, and fuelwoods (Windhoek: WWF (LIFE) Program).

LANGE, R. T. (1969), 'The piosphere: Sheep track and dung patterns', *Journal of Range Management*, 22: 396–400.

LEE, R. B. (1979), *The !Kung San: Men, Women and Work in a Foraging Society* (Cambridge: Cambridge University Press).

——(1984), *The Dobe Jul'hoansi: Case Studies in Cultural Anthropology*. (Toronto: Holt, Rinehart & Winston).

LEE, R. B., and DEVORE, I. (1976), *Kalahari Hunter-Gatherers* (Cambridge, Mass.: Harvard University Press).

LINDSAY, W. K. (1992), *Technical Assistance to the Project 'Initial Measures for the Conservation of the Kalahari Ecosystem'*, Final Report, Commission of the European Communities.

MCGOWAN INTERNATIONAL AND COOPERS LYBRAND (1988), *National Land Management and Livestock Project: Incentives/Disincentives Study*, I–III (Gaborone: Report prepared for the ministry of Agriculture).

MLGL (1991), 'A Monitoring Programme for the Settlements at Thankane, Kokotsha, Inalegolo, Monong, Ngwatlhe and Groot Laagte', Report prepared for Accelerated Remote Area Development Programme. Gaborone: MLG.

MORNA, C. L. (1979), 'Beyond the drought', *Africa Report*, 34/6: 30–3.

MORRISON, S. (1986), 'Dilemmas of sustaining parastatal success: The Botswana Meat Commission', *IDS Bulletin*, 17/1: 30–8 (Brighton: Institute of Development Studies).

MURRAY, M. (1988), *Management Plan for Central Kalahari and Khutse Game Reserves* (Gaborone: Kalahari Conservation Society).

ODA (1993), *Poverty in Botswana: A Status Report*, prepared by the Social Impact Assessment and Policy Analysis Corporation (Pty.) Ltd. (SIAPAC-Africa), Gaborone.

PARRIS, R., and CHILD, G. (1973), 'The importance of pans to wildlife in the Kalahari and the effect of human settlement on these areas', *Journal of the Southern African Wildlife Association*, 3/1: 1–8.

PATTERSON, L. (1987), 'Cordon fences', *Kalahari Conservation Society Newsletter*, 18: 10–11.

PERKINS, J. S. (1991), 'The Impact of Borehole Dependent Cattle Grazing on the Environment and Society of the Eastern Kalahari Sandveld, Central District, Botswana', Ph.D. Dissertation, University of Sheffield.

PERKINS, J. S. (1996), 'Botswana: Fencing out the equity issue. Cattleposts and cattle ranches in the Kalahari desert', *Journal of Arid Environments*, 33: 503–17.

——(1999), *Natural Resource Inventory and Monitoring Tool Kit Report for KD1 and NG33/34*, development of a Vegetation and Veld Product Monitoring Programme in the Controlled Hunting Areas KD1 and NG33/34 for USAID/ Department of Wildlife and National Parks, Gaborone.

PERKINS, J. S., and THOMAS, D. S. G. (1993*a*), 'Environmental responses and sensitivity to permanent cattle ranching in semi-arid western central Botswana', in D. S. G. Thomas and R. J. Allison (eds.), *Landscape Sensitivity* (Chichester: Wiley), 273–86.

——(1993*b*), 'Spreading deserts or spatially confined environmental impacts? Land degradation and cattle ranching in the Kalahari Desert of Botswana', *Land Degradation and Rehabilitation*, 4: 179–94.

POWELL, N. S. (1994), 'Participatory Land Use Planning: Methods Development Incorporating the Needs and Aspirations of Indigenous Peoples in Natural Resource Management; A Case from Eastern Bushmanland, Namibia', Report prepared for the Government of Namibia, Windhoek.

PRATT, D. J. (1967), 'A note on the overgrazing of burned grassland by wildlife', *East African Wildlife Journal*, 5: 178–9.

RPM (1995), *Ghanzi District Wildlife Management Area Management Plan: Matlho-a-Phuduhudu*, Draft Final Report to Ghanzi District Land Use Planning Unit, vol. i and ii. Prepared as part of the USAID Natural Resources Management Project.

SANDFORD, S. (1983), *Management of Pastoral Development in the Third World* (Chichester: John Wiley).

SILBERBAUER, G. B. (1981), *Hunter and Habitat in the Central Kalahari Game Reserve* (New York: Cambridge University Press).

SPINAGE, C. A., and MATLHARE, J. M. (1992), 'Is the Kalahari cornucopia fact or fiction? A predictive model', *Journal of Applied Ecology*, 29: 605–10.

STODDART, L. A., SMITH, A. D., and BOX, T. W. (1995), *Range Management* (New York: McGraw-Hill).

STORY, R. (1958), 'Some plants used by the Bushmen in obtaining food and water', *Botanical Survey of South Africa, Memoir No. 30* (Pretoria: Government Printers).

——(1964), 'Plant lore of the Bushmen', in D. H. S. Davis (ed.), *Ecological Studies in Southern Africa* (The Hague: Junk).

STUART-HILL, G. C., and PERKINS, J. S. (1997), 'Wildlife and Rangeland Management in the Proposed Nyae Nyae Conservancy', Report prepared for World Wide Fund for Nature- LIFE Project and Nyae Nyae Farmer's Co-operative.

SWEET, R. J. (1986), 'Boreholes in the Kalahari: The pros and cons', *Kalahari Conservation Society Newsletter*, 11: 10–11.

TANAKA, J. (1976), *The San: Hunter-Gatherers of the Kalahari: A Study in Ecological Anthropology*, trans. D. W. Hughes (Tokyo: University of Tokyo Press).

TAYLOR, R. T., and MARTIN, R. B. (1987), 'Effects of veterinary fences on wildlife conservation in Zimbabwe', *Environmental Management*, 11/3: 327–34.

VERLINDEN, A. (1994), *An Action Plan for the Management of Wildebeest Populations in the Kalahari* (Gaborone: Department of Wildlife and National Parks, Research Division).

WILLIAMSON, D., and WILLIAMSON, J. (1985*a*), 'Botswana's fences and the depletion of Kalahari wildlife', *Parks*, 10/2: 5–7.

——(1985*b*), *Kalahari Ungulate Movement Study* (Frankfurt: Frankfurt Zoological Society).

YOUNG, E., and ROSS, H. (1992), 'Using the Aboriginal rangelands: "Insider" realities and "outsider" perceptions'. *Rangeland Journal*, 16/2: 184–97.

5

Ecological Change in Kalahari Rangelands: Permanent or Reversible?

Andrew Dougill

Introduction

Agricultural intensification of domestic livestock production, especially on designated cattle ranches, throughout the Kalahari of Botswana has been experienced at an increasing rate in the last thirty years. Changes in land-use practices have led to a series of ecological changes that remain the focus of debates regarding the implications for continued sustainable agricultural development. This chapter analyses these debates and reviews recent studies undertaken to resolve key environmental uncertainties. In particular discussions aim to:

- Summarize the results of a range of environmental studies investigating ecological changes on Kalahari rangelands following the introduction of cattle grazing,
- Detail the spatial characteristics of ecological changes (at ranch- and micro-scales) which provide ecosystem resilience to permanent changes and land degradation,
- Outline environmental explanations of ecological changes in relation to determining factors of soil water and nutrient availability, direct herbivory impacts and changes in fire regimes,
- Present a conceptual model for ecosystem dynamics in Kalahari rangelands,
- Discuss the implications of environmental studies to the development of sustainable agricultural management strategies in the future Kalahari.

To achieve these aims, this chapter will draw on the full range of previous and ongoing environmental studies undertaken in the Makoba Ranch Blocks, Central District, Botswana (Fig. 5.1) since 1988. The information presented

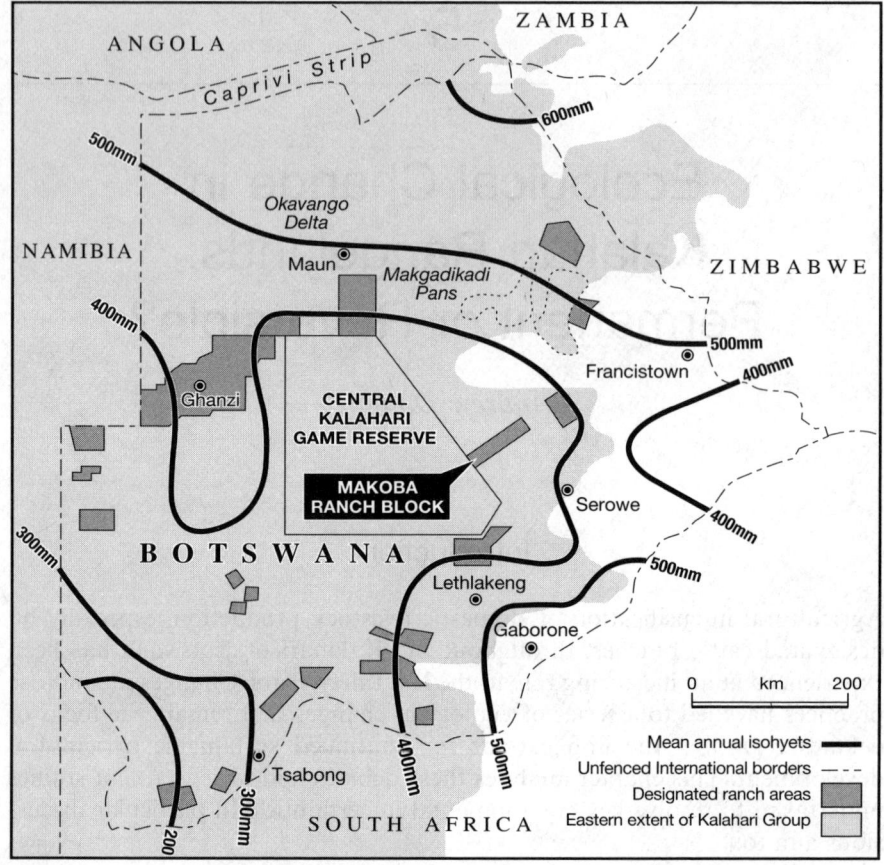

Fig. 5.1. TGLP ranch areas in Botswana.

represents the most detailed analysis of contemporary environmental
changes occurring on a specific area of the Kalahari. It synthesizes ranch-
scale ecological studies (Perkins and Thomas 1993*a*, *b*), micro-scale ecological studies (Dougill *et al.* 1998*b*; Dougill and Trodd 1999), ranch-scale
soil studies (Dougill and Cox 1995; Dougill *et al.* 1999), and micro-scale soil
studies (Dougill *et al.* 1998*a*), and summarizes ongoing remote sensing
studies detailing spatial patterns of ecological change (Trodd and Dougill
1998). Discussions in this chapter serve to highlight the environmental understanding that can be gained through a single site monitoring project carried
out over ten years. This informs a better understanding of changes in vegetation community composition and the natural resources upon which
pastoral-based rural livelihoods depend.

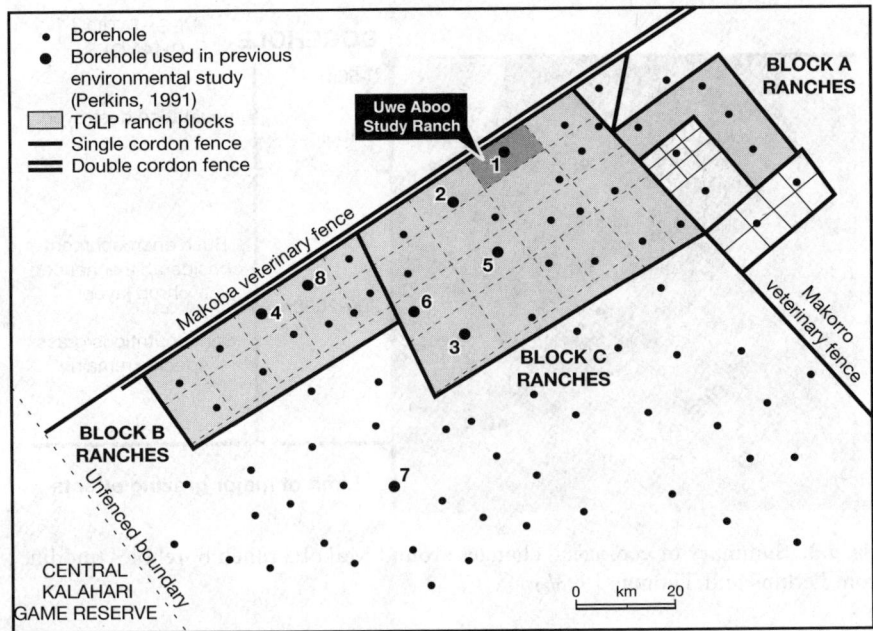

Fig. 5.2. Study ranches within the Makoba ranch block.

Makoba Ranches and Ranch-Scale Patterns of Ecological Change

The Makoba Ranch Block (Fig. 5.1) was chosen for these studies as an area typical of the structural adjustments occurring in the Kalahari following independence. It is one of twenty TGLP ranch blocks demarcated in 1975 and has experienced continued expansion in borehole provision since initial development. Consequently, the ranch block possesses a range of boreholes of different ages on which to analyse the ecological changes occurring as a result of the intensified cattle-rearing.

Ecological surveys around eight boreholes (Fig. 5.2) were initially undertaken during the 1988–9 wet season (Perkins 1991). Studies were based on a 'piosphere' approach (Georgiadis 1987) designed to investigate the link between the herbivore use intensity (HUI), which declines exponentially with distance away from the borehole, and changes in vegetation communities. Detailed vegetation analyses were undertaken for 25 m by 25 m quadrats at set distances of 0, 25, 50, 200, 400, 800, 1,500, 3,000, and 5,000 m from the borehole on each study ranch. In-depth analysis of these studies has been provided by Perkins and Thomas (1993*a, b*) and is simply summarized here. The most obvious and widespread ecological change recorded was the shift

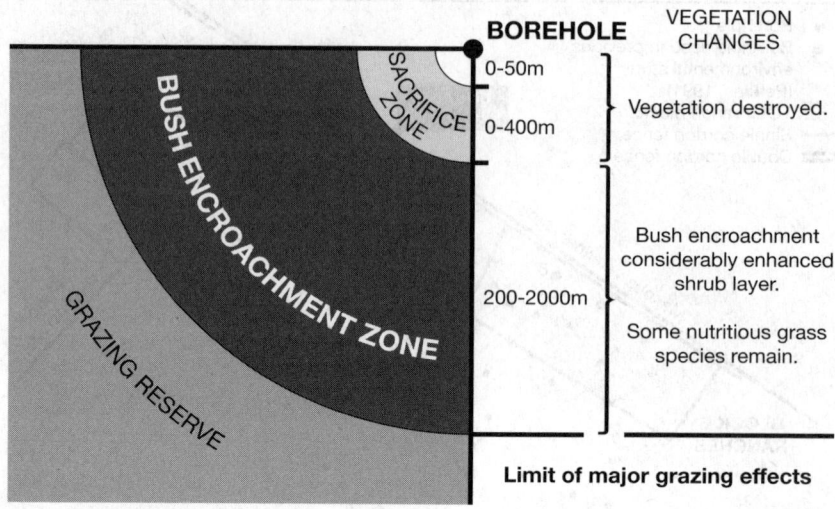

Fig. 5.3. Summary of ecological changes around Makoba ranch boreholes (modified from Perkins and Thomas 1993*b*).

from grass to bush dominance, a process termed 'bush encroachment', which has been noted in many other parts of the Kalahari (Cole and Brown 1976; Cooke 1983; Skarpe 1990; Ringrose *et al.* 1990, 1996). Perkins and Thomas (1993*a, b*) found that bush encroachment generally remains confined to an area within 2 km of the borehole (Fig. 5.3). Given the initial adoption of an '8-kilometre rule' for distances between boreholes within the TGLP, there remained, at this time, substantial areas largely unaffected by such ecological changes. These unencroached areas remained predominantly grass-dominated, and were termed the 'grazing reserve' (Perkins and Thomas 1993*a*). Field ecological evidence suggests that such a 'bush-encroached— grazing reserve' duality is the major ecological pattern on TGLP ranches. Consequently on a typical 6,400 ha TGLP ranch block, ecological and fodder diversity of both bush and grass remains within the ranch when only one central borehole is used to supply water to cattle.

Ecological heterogeneity in terms of a mixture of bush-dominant and grass-dominant areas has been clearly linked to sustainable pastoral production in semi-arid areas by Scoones (1995). This stresses the important role played by bush fodder in drought years. Such fodder can limit cattle losses and consequently retain a larger livestock base from which post-drought restocking can occur (Westoby, Walker, and Noy-Meir 1989; Behnke and Scoones 1993). Therefore it is not surprising that studies investigating the impacts of ecological changes on pastoral production figures have, to date, found no significant decline in pastoral production through time (Vossen

1990; White 1993) with variations linked solely to rainfall histories. This has been used to argue that Kalahari rangelands have not been subjected to major degradation (White 1993). However, Perkins and Thomas (1993b) also noted that bush encroachment is an expansive process with its greatest spatial extent occurring on the older boreholes. This, combined with the increasing development pressures that have caused a relaxation of the 8 km spacing between boreholes (Tsimako 1991), means that there is now both an ecological and management potential for the coalescence of bush-dominated areas. Coalescence would have a major impact on the fodder diversity of rangelands and therefore sustainable pastoral production, a factor now being recognized by farmers on the Makoba TGLP ranches.

Recent environmental studies have focused on one ranch, Uwe Aboo (Borehole 1 in Fig. 5.2), concentrating on the causal links between grazing, ecological changes, and the determining factors of soil water and nutrient availability. Findings are used here to present a conceptual model describing ecosystem dynamics in Kalahari rangelands, and to assess whether bush encroachment can be viewed as a permanent or reversible ecological change.

Ecological Theory and Research Implications

It is essential to understand wider debates concerning the functioning of semi-arid ecosystems and the sustainability of agricultural systems prior to analysing the findings of Kalahari studies. It is now acknowledged that variability far exceeds any equilibrial tendencies in semi-arid ecosystems (Ellis and Swift 1988; Westoby, Walker, and Noy-Meir 1989; Behnke and Scoones 1993). Weaknesses of the pastoral management strategies, such as fixed carrying capacities and constrained ranch blocks, adopted as a result of the acceptance of equilibrium theories are increasingly also being realized (Scoones *et al.* 1996; Warren 1995). The shift in ecological paradigm from equilibrium to non-equilibrium theories has also been accompanied by the closing of the divide between environmental science and socio-economic and cultural studies (Warren 1995). This shift importantly has led to the recognition that indigenous pastoral strategies are carefully adapted to the spatial and temporal variability which characterize semi-arid ecosystems (Scoones 1995).

The changes in ecological paradigm have many implications for environmental studies in the Kalahari. The paradigm shift can explain much of the confusion in the debates on whether ecological changes following grazing intensification represent land degradation (see Cooke 1983; Skarpe 1990; White 1993; de Queiroz 1993; Dougill and Cox 1995; Adams 1996 for a potted history of these lengthy debates). The principal confusion stems from the association between vegetation changes, such as bush encroachment, and land degradation. This direct association between vegetation *change* and land *degradation* can go against non-equilibrium ecological understanding, where rainfall variability plays a greater role in controlling plant growth than variations in grazing regimes (Ellis and Swift 1988; Friedel *et al.* 1993). Rainfall variability implies that vegetation changes measured at a given time are often reversible and therefore not necessarily indicative of degradation which (by definition) implies an 'effectively permanent' decline in the rate at which an ecosystem can support agricultural production (Abel and Blaikie 1989).

Studies on the Makoba ranches have progressed beyond the recognition of ecological changes to investigate the causes of the transition from grass-dominant to bush-dominant vegetation communities. Only through such investigation of the causes of ecological changes can questions be answered regarding the permanence of changes and therefore their significance in terms of land degradation and sustainability criteria.

Causes of Ecological Change in the Kalahari

Changes in semi-arid ecosystem structure (ratio of bush to grass cover) and ecological productivity are governed principally by soil water and soil nutrient availability, direct herbivory effects, and fire regimes (Scholes and Walker 1993; Belsky 1994). Studies after the initial ranch-scale vegetation surveys therefore focused on the link between cattle-grazing intensity and changes in these ecological determining factors. This has been achieved through detailed analysis of changes in soil hydrochemical characteristics and local-scale vegetation patterning studies to assess the relative effects of direct grazing impacts on vegetation, and the likely impacts on fire regimes of intensive grazing. As some direct management of herbivory pressures and fire regimes is possible, ecological changes can theoretically be classed as reversible provided soils have not changed greatly. Consequently, the recent shift in emphasis of land degradation studies to consider the impacts of management strategies on soil factors (e.g. Stocking 1995; UNEP 1997) is important, and provides a framework for the studies reported here.

Soil System Studies

To develop an understanding of the role assigned to soil water and nutrient availability in affecting ecosystem structure, a range of soil studies have been conducted at Uwe Aboo (Fig. 5.2). These studies have focused on assessing the applicability of the 'two-layer model of environmental change' proposed by Walker and Noy-Meir (1982), which has often been presented as the best explanation of bush encroachment in the Kalahari (e.g. Skarpe 1990, 1991; Perkins and Thomas 1993*a, b*). In this model, the balance between grass and

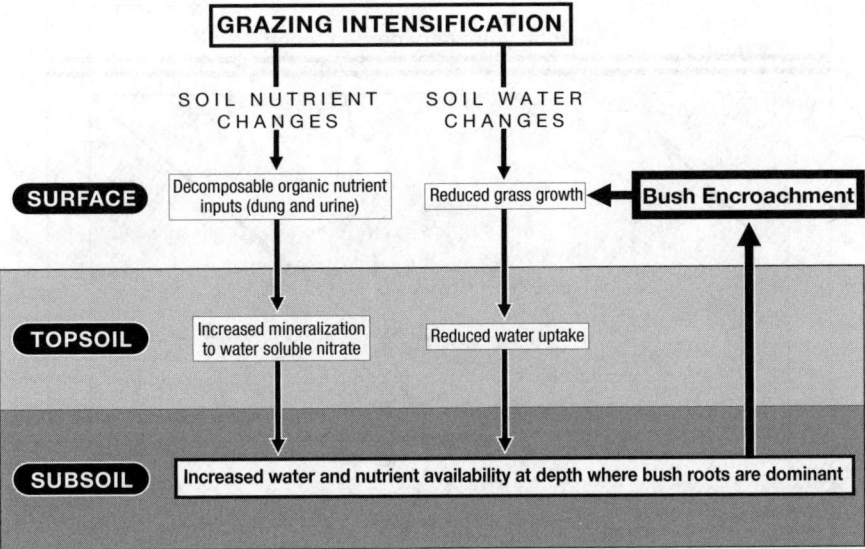

Fig. 5.4. Conceptual summary of the two-layer model of environmental change, as proposed by Walker and Noy-Meir (1982).

bush production is determined by the relative availability of soil water and nutrients (principally inorganic nitrogen and phosphorus) in different rooting zones. Grasses out-compete bushes for water and nutrients in the topsoil layer (0–0.5 m depth), while bushes have the competitive advantage in the subsoil (below 0.5 m depth) (Walker *et al.* 1981; Belsky 1990). According to the two-layer model, grazing changes this balance by suppressing grass growth and promoting soil water movement into the subsoil. At the same time, increased mineralization of organic nitrogen in cattle dung (which is more readily decomposed than residual plant litter) into nitrate (NO_3^-–N), enhances the leaching of this vital plant nutrient into the subsoil. Consequently, the two-layer model predicts that areas of intensive grazing experience significant increases in moisture content and inorganic nutrient concentrations in their subsoil, with bush encroachment being the inevitable (and therefore permanent) ecological consequence (Fig. 5.4).

The applicability of the proposed link between grazing intensity, ecological change, and changes in soil hydrochemical characteristics has been investigated by a ranch-scale monitoring programme and controlled small-scale experimental studies. Ranch-scale studies were carried out on a 3 km transect north of the 1973 borehole on Uwe Aboo Ranch (Fig. 5.5), with soils sampled at 25, 100, 400, 800, 1,600, and 2,800 m from the borehole and in an ungrazed grass-dominant control site at the neighbouring cordon fence. The analytical methodology followed and in-depth analysis of these studies has

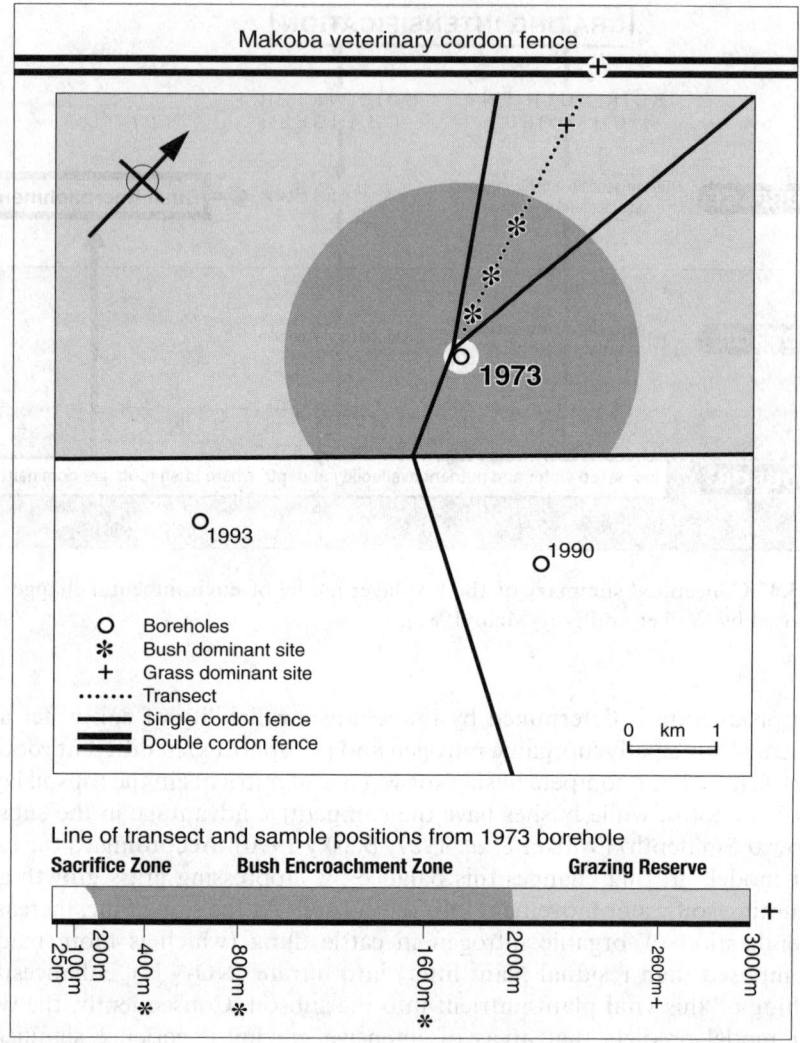

Fig. 5.5. Uwe Aboo ranch in 1993 showing study transect and location of bush-encroached and control study sites. General pattern of ecological change is marked according to Perkins and Thomas (1993*b*).

been outlined by Dougill and Cox (1995) and elaborated by Dougill *et al.* (1999). The controlled experiments are reported in detail by Dougill *et al.* (1998*a*). The results presented here provide an integrated summary of the soil-based research and their implications for discussions on the permanence of bush encroachment.

Ranch-scale studies highlight that for all the soil hydrochemical charac-

Table 5.1. Soil hydrological characteristics of intensively grazed bush-encroached, and ungrazed grass-dominant sites on the Uwe Aboo study ranch (Fig. 5.5)

Environmental Parameter	Bush-dominant	Grass-dominant	Implication
Hydraulic conductivity [$K(\theta)$]—rate of wetting front movement	12.8 ± 5.2 $n = 17$	12.3 ± 4.9 $n = 14$	No significant difference in rate of water movement
Field capacity (θ_{fc})— Topsoil moisture content (%) 30 hrs after 40 mm storm event	5.17 ± 0.18 $n = 8$	4.83 ± 0.43 $n = 8$	No significant difference in soil field capacity
Subsoil moisture content (%) 30 hrs after 40 mm storm event	1.10 ± 0.85 $n = 17$	0.99 ± 0.74 $n = 14$	No significant difference in soil water leached to the subsoil in initial wetting pulse
Subsoil moisture content (%) in 1993 dry season	0.90 ± 0.27 $n = 19$	0.93 ± 0.32 $n = 19$	No significant difference between residual subsoil moisture contents
Topsoil moisture content (%) in wet season	3.34 ± 0.22 $n = 4$	3.20 ± 0.03 $n = 4$	No significant difference
Subsoil moisture content (%) in wet season	4.54 ± 0.24 $n = 4$	4.13 ± 0.41 $n = 4$	No significant difference

Source: Data summarized from Dougill and Cox (1995) and Dougill, Heathwaite, and Thomas (1998*a*).

teristics investigated (soil moisture contents, and extractable NO_3^-–N, NH_4^+–N, and PO_4^{3-}–P concentrations in both topsoil and subsoil layers) significant changes are only experienced within a distance of 100 m from the borehole (Dougill and Cox 1995). Vitally, detailed comparisons between soil water and soil nutrient availability at an intensively grazed bush-encroached site and the ungrazed grass-dominant control site showed no significant differences in any of the soil hydrochemical characteristics investigated (Tables 5.1 and 5.2). These studies combine to provide strong evidence that intensive cattle grazing, and associated ecological changes, are not linked to (or caused by) changes in soil water and soil nutrient availability in Kalahari soils. These findings directly oppose the hypotheses proposed in the two-layer model (Fig. 5.4).

To explain the resilience to changes in soil water and nutrient availability patterns in Kalahari soils it is necessary to consider the processes affecting the transport and transformations of water and nutrients within soils. Soil water redistribution after rainfall events occurs along matrix flow pathways and is thus dependent on antecedent soil moisture conditions (Dougill *et al.* 1998*a*). Notably, the residually high subsoil moisture contents prevent

Andrew Dougill

Table 5.2. Soil inorganic nutrient availability at intensively grazed, bush-encroached, and ungrazed grass-dominant sites on the Uwe Aboo study ranch (Fig. 5.5)

Environmental Parameter	Bush-dominant	Grass-dominant	Implication
Topsoil extractable NO_3^--N concentration (mgN100g^{-1})	4.68 ± 2.53 n = 23	6.32 ± 1.98 n = 23	No significant difference
Subsoil extractable NO_3^--N concentration (mgN100g^{-1})	5.73 ± 3.81 n = 22	5.67 ± 1.06 n = 22	No significant difference
Topsoil extractable NH_4^+-N concentration (mgN100g^{-1})	5.13 ± 1.34 n = 23	5.07 ± 1.29 n = 23	No significant difference
Subsoil extractable NH_4^+-N concentration (mgN100g^{-1})	2.14 ± 0.88 n = 23	2.08 ± 1.21 n = 23	No significant difference
Topsoil extractable PO_4^{3-}-P concentration (mgP100g^{-1})	1.11 ± 0.44 n = 23	1.19 ± 0.61 n = 23	No significant difference
Subsoil extractable PO_4^{3-}-P concentration (mgP100g^{-1})	0.40 ± 0.22 n = 23	0.44 ± 0.27 n = 23	No significant difference

Source: Data summarized from Dougill and Cox (1995) and Dougill, Heathwaite, and Thomas (1998*a*).

Table 5.3. Grass biomass in sub-canopy and neighbouring open area quadrats for bushes on study transect in relation to their location in grazing intensity related vegetation zones, assigned by Perkins and Thomas (1993a, b) (see Fig. 5.5 for study locations)

Distance from bush rooting point (cm)	Bush-encroached sites	Grazing Reserve sites	Control sites
0–50	17.2 ± 5.2 (n = 48)	36.1 ± 17.1 (n = 14)	7.3 ± 22.4 (n = 30)
50–100	8.6 ± 4.5 (n = 27)	20.0 ± 7.5 (n = 9)	28.4 ± 39.7 (n = 29)
100–50	5.5 ± 6.3 (n = 6)	—	33.6 ± 49.0 (n = 26)
150–200	0.7 ± 1.3 (n = 3)	—	70.1 ± 74.8 (n = 14)
Open site	0.9 ± 0.6 (n = 48)	7.1 ± 5.7 (n = 14)	67.7 ± 53.6 (n = 30)

leaching losses beyond the top 1–2 m of soil profiles suggesting grass species remain able to compete successfully for all available soil water. Furthermore, process-based soil nutrient studies show that the infertile Kalahari soils experience very low rates of mineralization from organic (unavailable) to inorganic (available) forms of both nitrogen and phosphorus (Dougill *et al.* 1998*a*). These low mineralization rates allow synchrony between nutrient production and plant uptake, even in areas where this grass growth has been reduced through intensive grazing, which means that nutrient cycling remains topsoil dominated in both bush-dominant and grass-dominant settings. Consequently, soil studies suggest that ecological changes may be reversible, given the lack of significant changes in soil water and nutrient availability patterns, and that alternative models of environmental change are required. Such alternative models must focus on the other key determinants of ecological structure and productivity, namely the direct herbivory impacts of grazing and browsing and changes in fire regimes, and their impacts in relation to temporal patterns of rainfall variability.

Direct Herbivory and Fire Regime Effects: Local-Scale Ecological Studies

Investigations of the ecological effects of direct herbivory and fire regimes has typically only been possible through long-term ecological monitoring of vegetation community changes in relation to grazing, rainfall variability, and fire events (e.g. Scholes and Walker 1993). Long-term ecological monitoring has not been undertaken in the Kalahari, implying that although the link between ecological changes and grazing intensity has been proven (Perkins and Thomas 1993*a*, *b*) uncertainties remain in our understanding of the causes and timing of changes. Various alternative research angles are now available to attempt to reduce these uncertainties and have been applied in recent studies at Uwe Aboo. One possibility is to use remotely sensed satellite data to monitor semi-arid vegetation dynamics (Stafford-Smith and Pickup 1993). Initial spectral reflectance studies on the Makoba ranches (Trodd and Dougill 1998) have highlighted difficulties with the use of such data given the inability, for technical reasons, to differentiate between the reflectance characteristics of bush and grass canopies. Consequently, remote sensing studies in the Kalahari can, at present, provide us information only on temporal variations in the percentage and greenness of vegetation cover (Palmer and van Rooyen 1998; Richard and Poccard 1998), not the ratio of bush to grass cover. The use of temporal sequences of remotely sensed data to quantify changes in the variations of green biomass, measured using the surrogate of Normalized Difference Vegetation Index (NDVI), provides an opportunity to differentiate between bush- and grass-dominant areas based on their different growth response to rainfall.

Recent ecological debates have switched attention from the necessity for long-term monitoring in all semi-arid regions to the generalizations that can be drawn from shorter time-scale local-scale spatial patterns of bush and grass communities. For example, Jeltsch *et al.* (1996) conclude that increased grazing will lead to increased bush establishment and germination with bushes evenly distributed through an area, with spacing dependent on inter-tree competition for soil water and nutrients (Skarpe 1991). However, where occasional fires reverse bush encroachment, bushes will be patchily distrib-uted in thickets that exclude fire, and that bush encroachment will result from the spreading of such dense thickets. Ecological investigations at Uwe Aboo have thus switched to focus on the small-scale patterns of bush and grass cover on the study transect (Fig. 5.5). These show that there are fewer dis-crete bush clumps at encroached sites than in unencroached areas due to the greater number of bushes per clump and the larger average diameter of clumps (Dougill *et al.* 1998*b*). Bush encroachment appears to be occurring through the spreading of bush clumps to form denser bush thickets, with fire events representing a key mechanism preventing the immediate expansion of dense bush encroachment. It is proposed that with increased grazing, and therefore reduced herbaceous fuel loads in the dry season, the frequency, intensity, and spatial extent of fires will be reduced in intensively grazed areas, with expanding bush encroachment being the ecological consequence. Preliminary remote sensing studies of fire frequency carried out as part of the SAFARI project on a range of sites across the Kalahari in 1988 and 1992 (Justice *et al.* 1996) support this supposition. This hypothesized link to changes in fire regimes, whilst requiring further investigation, is important as it suggests that an increase in fires (whether natural or managed through prescribed burning) could return the ecosystem to a grass-dominant form. Local-scale bush pattern studies, therefore, favour the conclusion that bush encroachment on the Makoba ranches would be reversible if burning strat-egies were to be amended.

Bush expansion may even provide micro-environments that favour the per-sistence of grasses. Dougill *et al.* (1998*b*) found higher grass concentrations in sub-canopy locations than in open situations, in areas subject to grazing. In particular, dense and thorny encroaching *Acacia* species support dense grass biomass stands at bush-encroached sites (Dougill *et al.* 1998*b*). Such protected sub-canopy niches are typically characterized by increased soil water and nutrient availability (Schlesinger *et al.* 1990; Belsky *et al.* 1993), which explains the existence of nutritious grass species close to all the bore-holes studied in the Makoba ranches (Perkins and Thomas 1993*b*). The dense low-growing and thorny canopies of encroaching bushes provide a niche that maintains grass cover in intensively grazed sites and retains an extensive grass seed resource enabling rapid grass regrowth following rainfall, and also allows the transition back to grass-dominance to be re-established following any bush die-back. Ecological resilience is thus maintained within these rangeland systems. Bush die-back could occur through disturbances such as fires, drought, frost, lightning, wind, or wood harvesting, and as such could be either natural or managed, and would be followed by the transition back to ecological grass-dominance again implying that the ecological changes seen on the Makoba ranches remain reversible. Further research on the ecological causes of bush die-back and the potential for using chemical methods to control bush cover is vital to assessing the potential future sustainable management of Kalahari rangelands.

Conceptual Model of Ecosystem Dynamics: Permanent or Reversible Changes?

Findings from the full range of environmental studies on the Makoba ranches, and observations from elsewhere in the Kalahari, are used to provide a conceptual 'state and transition' model that summarizes the present understanding of ecosystem changes in Kalahari rangelands (Fig. 5.6). This model shows that currently on the Makoba ranches the lack of significant changes in the soil systems and the existence of protected niches that maintain significant grass cover in intensively grazed areas combine to highlight that ecological changes remain reversible. Natural environmental variability, due to rainfall variability and fire events, is important in providing ecosystem attributes that generate resilience and mitigate against ecological changes becoming immediately permanent. It is also clear, however, that cattle ranching does favour the rapid onset and expansion of bush encroachment. Even boreholes utilized for less than ten years have experienced such an ecological transition (Perkins and Thomas 1993*a*, *b*), and in time, these shifts are capable of producing extensive mature uniform stands of encroaching bush species. Evidence from other Kalahari rangelands, such as on communal lands surrounding the village of Lethlakeng (Fig. 5.1) which have been grazed intensively for over a hundred years, shows that over time bush encroachment can develop into stands of mature trees and bushes. These stands offer less protection from grazing resulting in the demise of the grass cover in the protected sub-canopy niche. Therefore, Fig. 5.6 portrays the bush encroachment on the Makoba ranches not as an 'effectively permanent' change, but as a stage in a chain of causation that may lead to quasi-permanent ecological change, or could still be reversed, depending on the environmental conditions experienced and the management strategies followed.

Ecological reversibility arguments should not be used to dismiss concerns about bush encroachment and future sustainable livestock production under a ranching framework. These concerns need to be viewed as real, given the assigned link between bush encroachment and decreased pastoral production in other areas, including neighbouring Namibia (Adams 1996), South Africa (Dean and McDonald 1994) and Sahelian Africa (Warren and Khogali 1992). The increasing density of boreholes and the fragmentation of pastoral management units, both of which have occurred extensively in recent years on the Makoba ranches, pose threats to sustainable productivity even without permanent ecological change. As management units decrease in size, the

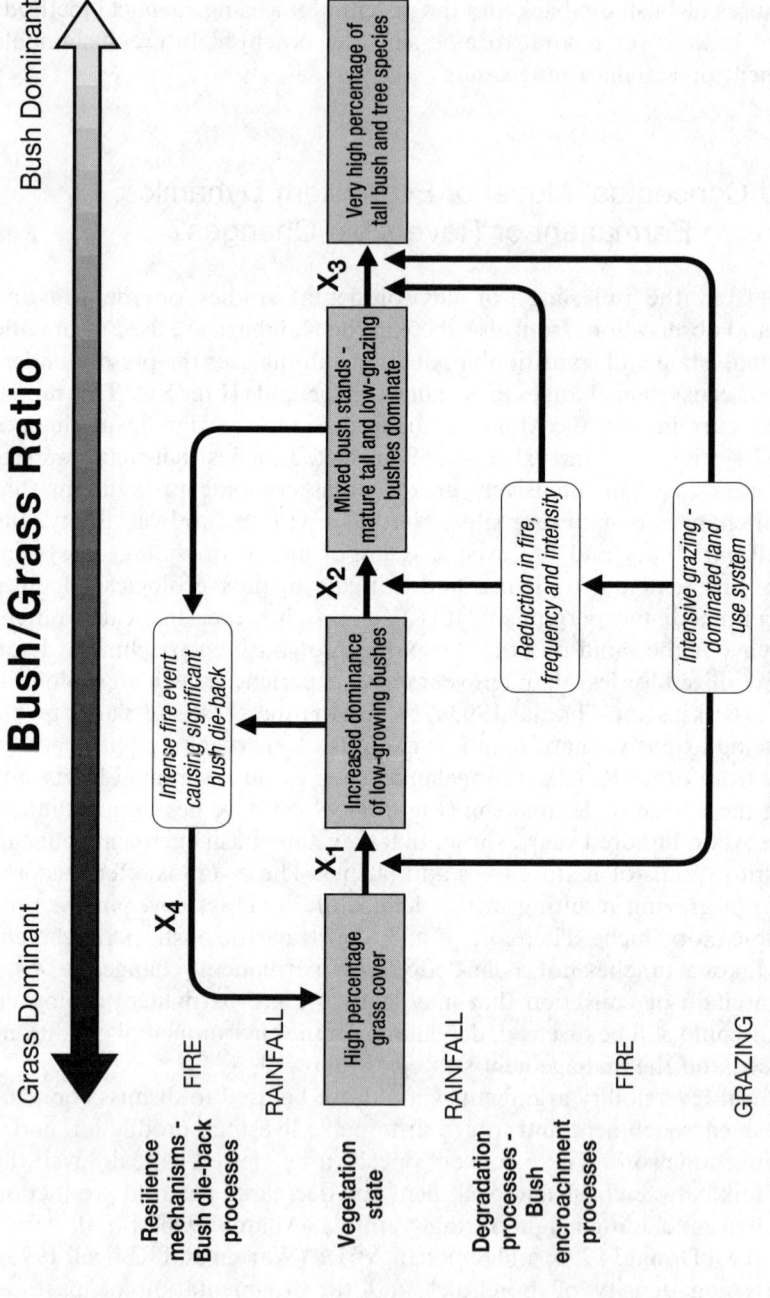

Fig. 5.6. Conceptual model of the factors affecting ecological change in Kalahari rangelands.

potential for them to become almost totally bush-dominated increases, as bush zones merge to form large areas of relatively unproductive rangeland. Under these circumstances, the evidence provided here suggests that it is not permanent ecological changes that will lower productivity, as the soil system and the ecosystem's potential to support grasses will remain unaltered. It is the modifications in the pastoral management systems, and the fencing of ranch units, that will be responsible for reduced livestock yields.

Implications for Sustainable Agricultural Development

The suite of environmental studies describing bush encroachment as an inevitable, but plausibly reversible, consequence of cattle-ranching (Fig. 5.6) have a number of important implications for the sustainable management of Kalahari rangelands. Sustainable management is dependent on the maintenance of the mix of bush-dominant and grass-dominant areas. However, certain management factors must be considered to prevent the threat of a transition to extensive, merged areas of bush-encroached rangeland.

First, the importance of the residual dry season grass cover to recovery following rainfall and other major disturbances means there is a need to reduce the defoliating effects of cattle during and immediately after droughts, when grasses are most susceptible to grazing-induced mortality (Mott *et al.* 1992). There is a need to ensure that over-utilization of drought-affected rangeland does not occur. Consequently, on a given ranch a distribution of cattle that matches the spatial availability of grass forage appears logical, rather than the intensive use of paddocks to the extent that nearly all grass cover is removed. This is especially true as the resting of other paddocks leads to widespread senescence of ungrazed grasses that then reduces grass regrowth in following wet seasons, a problem noted on Uwe Aboo since 1995 when strict rotational grazing strategies were introduced. The long-term resilience of grass communities could be maintained through rapid destocking programmes that ensure livestock owners gain an incentive to destock (or move cattle to wetter areas) at drought onset and have access to livestock capital in the post-drought period to enable rapid restocking (Toulmin 1995). Cattle mortality episodes that reduce herd sizes in drought events (White 1993) should not be completely prevented. Alternatively, opportunistic tracking strategies (Scoones 1995) could be set up that allow the movement of livestock to areas of adequate grass supply, making the most of the regional-scale spatial ecological variability caused by rainfall variations. Such movements do occur on cattle-trucks between different ranches owned by the same person (Kabelo, pers. comm.), but they have not reached the sophisticated level seen in some other semi-arid areas (e.g. Pearce 1995). The increased mobility option would allow rapid restocking in wet years following droughts, but would require co-operation between ranch-owners on a scale not presently seen in Botswana. The development of Farmer Groups within the area offers an opportunity whereby such social capital improvements could be used to improve the sustainability of natural resource use. These opportunities demonstrate clearly the need to consider future agricultural use of rangelands using such an integrated livelihoods approach.

Secondly, limits need to be imposed on both the density of boreholes and the continued fragmentation of ranch blocks into smaller paddocks which disturbs the bush-encroached–grazing reserve duality on ranches (Perkins and Thomas 1993a). At the very least, the cost implications of borehole-drilling and fencing operations, the losses that result from decreased heterogeneity of fodder resources, and costs of mitigating control measures (in terms of bush clearance) need to be added into the decision-making processes in the region. It is essential to use the increasing evidence from other areas to show that increasing borehole density does not translate directly to increased profitability.

Finally, the model of ecosystem dynamics (Fig. 5.6) implies that the role of major disturbances, especially fire events, could be vital in reversing the process of expanding bush encroachment. Further studies are required to assess the applicability of this link and the competitive dominance provided for grasses following fires on Kalahari rangelands. These could be developed to examine the possibility of incorporating prescribed burning practices and/or bush clearance programmes in management strategies designed to maintain grass-dominant areas, allowing continued, sustainable, intensive cattle ranching. Consequently, the need to continue long-term ecological monitoring of vegetation changes and grazing levels remains a priority for Kalahari rangelands. The wealth of studies conducted on the Makoba ranches and the similarities here to land-use changes elsewhere on Kalahari rangelands implies they provide an ideal study site for such integrated ground-based and remotely sensed monitoring.

Conclusions

Environmental studies on the Makoba ranches have characterized bush encroachment as the major ecological change associated with the intensification of cattle grazing over the last thirty years. However, the findings reviewed here contest the immediate attribution of this change as effectively permanent, which is inherent in previous classifications of bush encroachment as land degradation. It is proposed that the lack of significant long-term declines in livestock production (Vossen 1990; White 1993) can be explained by ranch-scale and local-scale patterns of ecological change up to the time of these studies. On a ranch scale, the restriction of bush encroachment to within a 2-km radius of the borehole (Fig. 5.3) implies that, with the adoption of 8 km spacing between boreholes under the initial TGLP guide-lines, fodder diversity was maintained on ranches through a mix of bush-dominant and grass-dominant areas. Micro-scale studies demonstrate that the continued expansion of bush encroachment is checked by resilience mech-anisms linked to environmental variability (drought and fire events), and the existence of fertile protected sub-canopy niches, which maintain a significant grass cover allowing rapid grass regrowth following rainfall and/or major dis-turbances, such as fire events.

The studies reported here show that bush encroachment has occurred inde-pendently of soil hydrochemical characteristics and therefore may be reversed by interactions between pastoral management strategies and droughts and fires. Consequently, the threat to sustainable livestock production in the years ahead is more likely to occur as a function of changes in management practices, notably the establishment of more borehole watering-points and the fragmentation of management units, rather than as a function of per-manent ecological changes.

References

ABEL, N. O. J., and BLAIKIE, P. M. (1989), 'Land degradation, stocking rates and con-servation policies in the communal rangelands of Botswana and Zimbabwe', *Land Degradation and Rehabilitation*, 1: 101–23.

ADAMS, M. (1996), 'When is ecosystem change land degradation? Comments on Land degradation and grazing in the Kalahari', *Pastoral Development Network Paper*, 39e (London: Overseas Development Institute).

BEHNKE, R. H., and SCOONES, I. (1993), 'Rethinking range ecology: Implications for rangeland management in Africa', in R. H. Behnke, I. Scoones, and C. Kerven (eds.), *Range Ecology at Disequilibrium: New Models of Natural Variability and Pas-toral Adaptation in African Savannas* (London: Overseas Development Institute), 1–30.

BELSKY, A. J. (1990), 'Tree/grass ratios in East African savannas: A comparison of existing models', *Journal of Biogeography*, 17/4: 483–9.

——(1994), 'Influences of trees on savanna productivity: Tests of shade, nutrients and tree-grass competition', *Ecology*, 75: 922–32.

BELSKY, A. J., MWONGA, S. M., AMUNDSON, R. G., DUXBURY, J. M., and ALI, A. R. (1993), 'Comparative effects of isolated trees on their undercanopy environments in high-rainfall and low-rainfall savannas', *Journal of Applied Ecology*, 30: 143–55.

COLE, M. M., and BROWN, R. C. (1976), 'The vegetation of the Ghanzi area of western Botswana', *Journal of Biogeography*, 3: 169–96.

COOKE, H. J. (1983), 'The struggle against environmental degradation—Botswana's experience', *Desertification Control Bulletin*, 8: 9–15.

DEAN, W. R. J., and McDONALD, I. A. W. (1994), 'Historical changes in stocking rates of domestic livestock as a measure of semi-arid and arid rangeland degradation in the Cape Province, South Africa', *Journal of Arid Environments*, 26: 281–96.

DE QUEIROZ, J. S. (1993), 'Range degradation in Botswana: Myth or reality?' *Pastoral Development Network Paper* 35b (London: Overseas Development Institute).

DOUGILL, A. J., and COX, J. (1995), 'Land degradation and grazing in the Kalahari: New analysis and alternative perspectives', *Pastoral Development Network Paper 38c* (London: Overseas Development Institute).

DOUGILL, A. J., and TRODD, N. M. (1999), 'Monitoring and modelling open savannas using multisource information: Analyses of Kalahari studies', *Global Ecology and Biogeography*, 8: 211–21.

DOUGILL, A. J., HEATHWAITE, A. L., and THOMAS, D. S. G. (1998a), 'Soil water movement and nutrient cycling in semi-arid rangeland: Vegetation change and system resilience', *Hydrological Processes*, 12: 443–59.

DOUGILL, A. J., THOMAS, D. S. G., and HEATHWAITE, A. L. (1999), 'Environmental change in the Kalahari: Integrated land degradation studies for non equilibrium dryland environments', *Annals, Association of American Geographers*, 89: 420–42.

DOUGILL, A. J., TRODD, N. M., and SHAW, M. J. (1998b), 'Spatial patterns of grass biomass in a grazed semi-arid savanna: Ecological implications', *The North West Geographer*, 2: 41–52.

ELLIS, J. E., and SWIFT, D. M. (1988), 'Stability of African pastoral ecosystems: Alternate paradigms and implications for development', *Journal of Range Management*, 41: 450–9.

FRIEDEL, M. H., PICKUP, G., and NELSON, D. J. (1993), 'The interpretation of vegetation change in a spatially and temporally diverse arid Australian landscape', *Journal of Arid Environments*, 24: 241–60.

GEORGIADIS, N. J. (1987), 'Responses of savanna grasslands to extreme use by pastoralist livestock', Ph.D. Thesis, Syracuse University, New York.

JELTSCH, F., MILTON, S. J., DEAN, W. R. J., and VAN ROOYEN, N. (1996), 'Tree spacing and coexistence in semi-arid savannas', *Journal of Ecology*, 84: 583–95.

JUSTICE, C. O., KENDALL, J. D., DOWTY, P. R., and SCHOLES, R. J. (1996), 'Satellite remote sensing of fires during the SAFARI campaign using NOAA advanced very high resolution radiometer data', *Journal of Geophysical Research*, 101/23: 851–63.

MOTT, J. J., LUDLOW, M. M., RICHARDS, J. H., and PARSON, A. D. (1992), 'Causes of variation in seasonal response to defoliation in three tropical savanna grasses', *Australian Journal of Agricultural Research*, 43: 241–60.

PALMER, A. R., and VAN ROOYEN, A. F. (1998), 'Detecting vegetation change in the southern Kalahari using Landsat TM data', *Journal of Arid Environments*, 39: 143–53.

PEARCE, F. (1995), 'Shepherds or Wise Men?' *New Scientist*, 24 Dec. 1995, 24–7.

PERKINS, J. S. (1991), 'The Impact of Borehole Dependent Cattle Grazing on the Environment and Society of the Eastern Kalahari Sandveld, Central District, Botswana', Ph.D. thesis, University of Sheffield.

PERKINS, J. S., and THOMAS, D. S. G. (1993a), 'Environmental responses and sensitivity to permanent cattle ranching in semi-arid western central Botswana', in D. S. G. Thomas and R. J. Allison (eds.), *Landscape Sensitivity* (Chichester: Wiley), 273–86.

——(1993b), 'Spreading deserts or spatially confined environmental impacts? Land degradation and cattle ranching in the Kalahari desert of Botswana', *Land Degradation and Rehabilitation*, 4: 179–94.

RICHARD, Y., and POCCARD, I. (1998), 'A statistical study of NDVI sensitivity to seasonal and interannual rainfall variations in Southern Africa', *International Journal of Remote Sensing*, 19: 2907–20.

RINGROSE, S., CHANDA, R., NKAMBWE, M., and SEFE, F. (1996), 'Environmental change in the mid-Boteti area of North-Central Botswana: Biophysical processes and human perceptions', *Environmental Management*, 20: 397–410.

RINGROSE, S., MATHESON, W., TEMPEST, F., and BOYLE, T. (1990), 'The development and causes of range degradation features in south-east Botswana using multi-temporal Landsat MSS imagery', *Photogrammetric Engineering and Remote Sensing*, 56: 1252–62.

SCHLESINGER, W. H., REYNOLDS, J. F., CUNNINGHAM, G. L., HUENNEKE, L. F., JARRELL, W. M., VIRGINIA, R. A., and WHITFORD, W. G. (1990), 'Biological feedbacks in global desertification', *Science*, 247: 1043–8.

SCHOLES, R. J., and WALKER, B. H. (1993), *An African Savanna: Synthesis of the Nylsvley study* (Cambridge: Cambridge University Press).

SCOONES, I. (1995), 'New directions in pastoral development in Africa', in I. Scoones (ed.), *Living with Uncertainty: New Directions in Pastoral Development in Africa* (London: Intermediate Technology Publications), 1–36.

SCOONES, I., CHIBUDU, C., CHIKURU, S., JERANYAMA, P., MACHAKA, D., MACHANJA, W., MAVEDZENGE, B., MOMBESHORA, B., MUDHARA, M., MUDZIWO, C., MURIMBARIMBA, F., and ZIZEREZA, B. (1996), *Hazards and Opportunities: Farming Livelihoods in Dryland Africa, Lessons from Zimbabwe* (London: ZED Books).

SKARPE, C. (1990), 'Shrub layer dynamics under different herbivore densities in an arid savanna, Botswana', *Journal of Applied Ecology*, 27: 873–85.

——(1991), 'Spatial patterns and dynamics of woody vegetation in an arid savanna', *Journal of Vegetation Science*, 2: 565–72.

STAFFORD-SMITH, M., and PICKUP, G. (1993), 'Out of Africa, looking in: Understanding vegetation change', in R. H. Behnke, I. Scoones, and C. Kerven (eds.), *Range Ecology at Disequilibrium: New Models of Natural Variability and Pastoral Adaptation in African Savannas* (London: Overseas Development Institute), 196–226.

STOCKING, M. A. (1995), 'Soil erosion and land degradation', in T. O'Riordan (ed.), *Environmental Science for Environmental Management* (London: Longman), 223–42.

TOULMIN, C. (1995), 'Tracking through drought: Options for destocking and restocking', in I. Scoones (ed.), *Living with Uncertainty: New Directions in Pastoral Development in Africa* (London: Intermediate Technology Publications), 95–115.

TRODD, N. M., and DOUGILL, A. J. (1998), 'Monitoring vegetation dynamics in semi-arid African rangelands: Use and limitations of Earth observation data to characterize vegetation structure', *Applied Geography*, 18/3: 15–33.

TSIMAKO, B. (1991), *The Tribal Grazing Land Policy (TGLP) Ranches: Performance to Date* (Gaborone: Ministry of Agriculture).

UNEP (1997), *World Atlas of Desertification* (Sevenoaks: Edward Arnold).

VOSSEN, P. (1990), 'Algorithm for the simulation of bare sandy soil evaporation and its application for the assessment of planted areas in Botswana', *Agricultural and Forest Meteorology*, 30: 173–88.

WALKER, B. H., and NOY-MEIR, I. (1982), 'Aspects of the stability and resilience of savanna ecosystems', in B. J. Huntley and B. H. Walker (eds.), *Ecology of Tropical Savannas* (Berlin: Springer-Verlag), 556–90.

WALKER, B. H., LUDWIG, D., HOLLING, C. S., and PETERMAN, R. S. (1981), 'Stability of semi-arid savanna grazing systems', *Journal of Ecology*, 69: 473–98.

WARREN, A. (1995), 'Changing understandings of African pastoralism and the nature of environmental paradigms', *Transactions, Institute of British Geographers*, 20: 193–203.

WARREN, A., and KHOGALI, M. (1992), *Assessment of Desertification and Drought in the Sudano-Sahelian Region 1985–1991* (New York: United Nations Sudano-Sahelian Office).

WESTOBY, M., WALKER, B. H., and NOY-MEIR, I. (1989), 'Opportunistic management for rangelands not at equilibrium', *Journal of Range Management*, 42: 266–74.

WHITE, R. (1993), *Livestock Development and Pastoral Production on Communal Rangeland in Botswana* (Gaborone: Botswana Society).

6

Environmental Change, Entitlements, and Poverty in Pastoral Systems

Deborah Sporton and David S. G. Thomas

Introduction

This chapter examines society–environment links arising from the expansion of commercial pastoralism into the dry-subhumid to arid Kalahari Desert of Botswana. The context of structural and land-use changes in Botswana is provided by the 1975 Tribal Grazing Land Policy (TGLP) which aimed to enhance national economic and social development by reforming the organization of Botswana's livestock industry, improving the livelihoods of rural dwellers, and ameliorating environmental degradation.

TGLP has been investigated and criticized in terms of its unjustified underpinnings (Sandford 1980); disadvantageous impacts on particular sectors of the rural population (e.g. Hitchcock 1980); contributions to wildlife population declines (Williamson and Williamson 1985); assumptions about available empty lands (Cooke 1985); and poor implementation and limited benefits for the livestock industry (Tsimako 1991; Hitchcock 1978). While these studies, many conducted over a decade ago, provide valuable critiques of the policy, they have not integrated environmental and social assessments, examined the spatial variability of the policy's impacts, or explored the ways in which rural livelihoods are responsive to environmental variability and the ways in which interactions have been altered under structural land-use change.

Understanding Social and Environmental Change

In the social sciences, research into human dimensions of land degradation has frequently highlighted the detrimental influences of national and international structural adjustment policies on post-colonial society–environment relationships (Mearns 1991; Redclift 1995; Killick 1991, 1995). Croll and Parkin (1992) have shown how Western constructions of the environment have dominated past investigations, ignoring the contributions of indigenous knowledge systems and experience. Today, political ecology (Blaikie and Brookfield 1987), post-modern (Gandy 1996) and post-structuralist (Peet and Watts 1996) critiques increasingly place emphasis on local knowledge and the achievement of sustainable development through the empowerment of grass-roots communities and the recognition of the complexity and variability of enviro-social relations. In this context the concept of environmental entitle-ments (Leach and Mearns 1991), namely the resources available to people and their abilities to manage them, potentially offers useful insights into the study of environmental degradation and livelihoods but has yet to be widely applied in the field.

Significant developments in understanding dryland environmental dynam-ics have also occurred during the last decade, especially in the context of pas-toral systems. Equilibrium models, often associating range degradation with traditional pastoralism, have been challenged within the context of disequi-librium theory (Behnke *et al.* 1993). Savanna systems are seen as complex, dynamic matrices of plant, soil, climate, and land-use variables, often well adapted to drought and other disturbances (Walker 1985). The spatial dimen-sion of environmental change in utilized systems is now better understood too (e.g. Scoones 1992; Perkins and Thomas 1993*a*; Dougill *et al.* 1999; see Ch. 5 this volume).

Improved understandings of the environmental dimension are critical to investigations of social aspects of rural change where communities are closely allied to the land. Understanding the nature of, and susceptibility to, vari-ability and disturbance in such environments has a further relevance given predictions of enhanced drought frequencies and moisture deficits during the twenty-first century under the impact of global warming (Williams and Balling 1995). As Tiffen *et al.* (1994: 17) have argued, 'unproven generalisa-tion has always been an enemy of deeper understanding of Africa's envi-ronmental problems'. Our thesis has been that it is advantageous, given the complex interplay of dynamic environmental and social variables, to investi-gate both physical and human components simultaneously and in a manner that permits integrated analysis.

The Tribal Grazing Land Policy

In Botswana the development of commercial livestock ranches intensified during the 1970s with the Tribal Grazing Land Policy (1975), which aimed to enhance economic and social development by reforming the organization of Botswana's livestock industry. Agricultural activities in Botswana had hitherto been closely associated with, environmental opportunities, ethnic background, and the burgeoning influence of élites. In rural areas, mixed activities such as herding and foraging have traditionally predominated and are practised according to ethnic background and environmental opportunities (Campbell 1986; Hitchcock 1985). The harsh environment of the Kalahari, which occupies much of the centre and west of the country, precludes cultivation which is largely practised to the east where the country's population is also concentrated. The significance of TGLP is that it has resulted in the expansion of sedentary borehole-centred livestock production into the environmentally marginal Kalahari and away from the wetter east where 70 per cent of the national herd was previously located. It was hoped that environmental problems perceived to be due to overstocking in the east and social problems arising from the domination of large cattle-owners and widening inequalities between rich and poor would be alleviated if larger herds were moved away from the east, allowing smaller farmers to improve their livelihoods.

As a result of the policy, fenced, borehole-centred leasehold ranches were allocated, of which 218 had been established by 1984 (Tsimako 1991). Those eligible to lease ranches from the state were cattle-owners or members of syndicates who owned at least 400 head of cattle and who would then have sole water rights and de facto control over grazing and other resources on the ranch. The Tribal Grazing Land Policy was, however, founded upon a number of false assumptions. The Kalahari was represented, for example, as a vast untapped grazing resource; an assumption dating back to the colonial period (Debenham 1952) which ignored the fact that many of the areas zoned under TGLP were already inhabited and used not only for hunting and gathering but also for seasonal grazing. Indeed, cattleposts had been operative, in some places, for over 200 years (Campbell, Main, and Associates 1991) and were associated with traditional opportunistic grazing and transhumance practises which helped to spread risks in times of drought (Moyo *et al.* 1991; White 1993). Thirdly, large cattle-owners were not willing to relinquish access to communal lands for grazing as envisaged and continued to use their dual grazing rights thereby further marginalizing smaller farmers (Peters 1994). One of the stated objectives of TGLP was to improve the social and economic well-being of rural populations through service centres which were created to absorb displaced populations, otherwise known as 'Remote Area Dwellers' (RADs) and provide education and health facilities, and to make provision for alternative employment opportunities. The term 'Remote Area Dweller' or RAD refers to the rural poor living outside gazetted villages. The Setswana term for RAD is *tengyanateng*, literally 'the farthest people', or, as it is sometimes translated, those from 'the deep within the deep' (Mogwe 1992). In establishing Service Centres it was hoped to reduce levels of social and economic inequality believed to be associated with traditional pastoralism.

In particular, these service centres were the loci for labour-based drought relief, feeding programmes, and also Remote Area Development Initiatives.

The development of Botswana's pastoral economy has therefore been closely associated with ecological and socio-economic tensions (Braat and Opschoor 1990). Against the background of, more or less, continuous drought since the 1980s and despite a series of consultancy reports fore-warning of some of the deleterious social consequences of ranching (Campbell, Main, and Associates 1991; Hitchcock 1985; McGowan 1988; Sandford 1980; Tsimako 1991), a National Policy on Agricultural Develop-ment was passed in 1991 which called for further increases in livestock pro-ductivity through the fencing of communal lands (Republic of Botswana 1991). As borehole-owners are granted exclusive grazing rights, there is potential for other users to be reduced to the status of squatters (Good 1992) and for the distribution of wealth to become increasingly polarized, for example, it has been estimated that 7 per cent of the population own half the national herd while a further 45 per cent own no cattle at all. It is feared that piecemeal enclosure of the type advocated in the 1991 policy will progres-sively concentrate the rural poor on the remaining communal lands with severe social and environmental consequences (Peters 1994).

Botswana is therefore currently proceeding along a path designed to maximize foreign earnings through diamond and beef exports (Ministry of Finance and Development 1991), and is arguably a classic example of a bonanza economy. The government has persevered with the ranch model although recent evidence suggests that a more flexible system of range management *may* prevent degradation in a marginal environment such as the Kalahari. This chapter develops previous research to explore the ways in which rural livelihoods are responsive to environmental variability and the ways in which these interactions have been altered under the Tribal Grazing Land Policy.

Research Design

The research design incorporated three ranch blocks designated and imple-mented under the 1975 TGLP and located in the vicinity of Serowe (Study Area 1), Ncojane (Study Area 2), and Tsabong (Study Area 3). These areas are located along the south-west (driest) to north-east (wettest) climate gra-dient (Fig. 6.1), which also represents a gradient of distance from the most densely populated eastern hardveld area of the country and therefore access to other livelihood opportunities.

Data were collected between 1994 and 1996 using an integrated multi-method research strategy combining extensive environmental and social surveys with in-depth ethnographic methods (for more detail see Thomas and Sporton 1997; Twyman *et al.* 1999; Thomas *et al.* 2000). In the absence of suitable baseline data, an extensive questionnaire survey of forty-one ranches

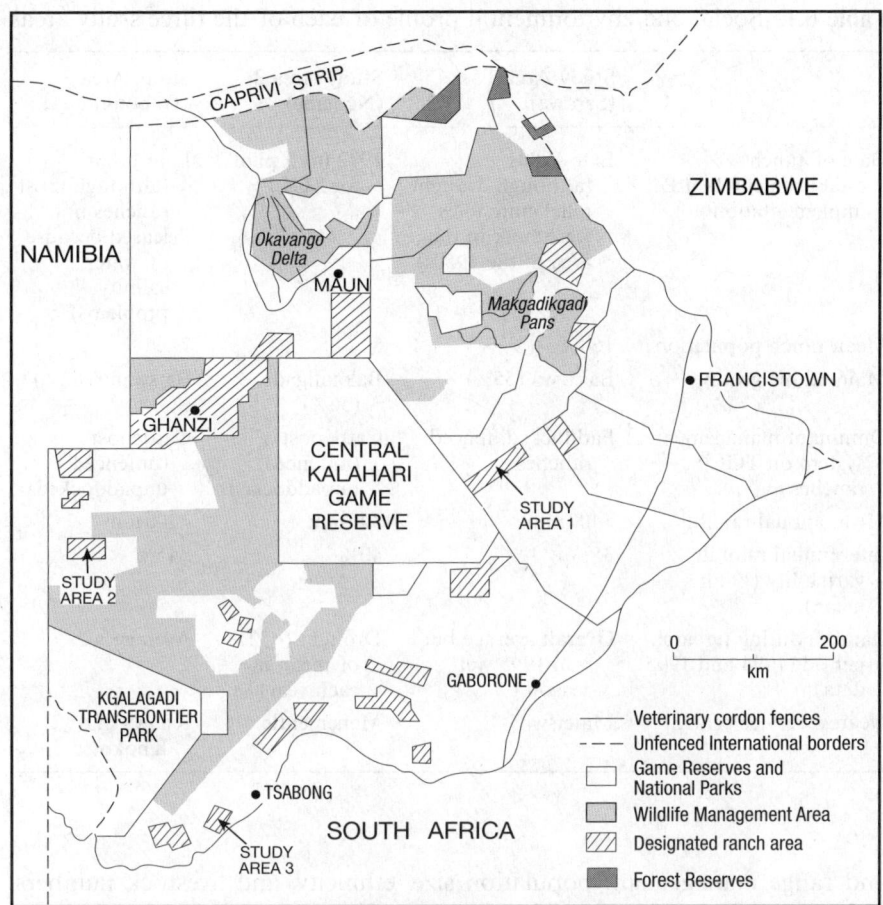

Fig. 6.1. Location of the three study areas.

representing each of the three study areas was conducted in which all residents, a total of 283 people, were included. To provide an environmental context for the social component an ecological survey was conducted on ranches in each of the three blocks to establish ecological opportunities and levels of degradation. Techniques including numerical gradient and classification techniques (e.g. CANOCO and TWINSPAN: applied in a comparable environmental study), were used to analyse the vegetation data to provide a better understanding of spatial differences within and between ranch areas and the principal causes of these differences. On the basis of the findings of the questionnaire and ecological surveys a number of ranches were selected for further in-depth investigation in the form of case studies. In particular, ranches were chosen to reflect variability in level of lessee involvement

Table 6.1. Social and environmental profile of each of the three study areas

	Study Area 1 (Serowe)	Study Area 2 (Ncojane)	Study Area 3 (Tsabong)
Date of ranch establishment/TGLP implementation	Late 1980s (although drought relief boreholes were sunk in the area in the 1950s)	1972 (as a pilot for TGLP)	Late 1980s (although most ranches not leased because of water salinity problems)
Mean ranch population	16	8	7
Main ethnic group	Basarwa (55%)	Bakgaligadi (56%)	Batswana (82%)
Dominant management system on TGLP ranches	Paddocked, fenced ranches	Cattlepost (unfenced, unpaddocked)	Cattlepost (unfenced, unpaddocked)
Mean annual rainfall	>400 mm	300 mm	<200 mm
Interannual rainfall variability (% of mean)	35%	40%	>45%
Rainfall during research period (1994 and 1995 data)	Overall average but > in 1995 wet season	Drought (<50% of mean in each year)	Average
Nearest service centre	Maletswai	Metsimentle	Khawa and Khokotsa

and range management, population size, ethnicity, and livestock numbers. Using similar agendas, interviews were conducted with governmental and non-governmental officials in Gaborone; with district officers and local NGO representatives in Serowe (Study Area 1), Ncojane (Study Area 2), and Tsabong (Study Area 3); and with ranch lessees, managers, and herders. Indicators of the key social and environmental characteristics of each of the three areas are summarized in Table 6.1.

Population, Environment, and Land-Use Change

TGLP, Population Displacement, and Livelihood Strategies

Settlement in the Kalahari has traditionally been focused around boreholes, yet the stated objectives of the TGLP to privatize the range has threatened settlement, mobility, and livelihood systems. In the initial stages of TGLP, with the zoning of the ranch blocks, large numbers of people were displaced from the ranch areas either onto communal lands or into newly created

Table 6.2. Population change in TGLP ranch blocks

	Previous Population Size	1994 Population Size	Percentage Change
Study Area 1	(1988) 285[a]	199	−30%
Study Area 2	(1991) 716[b]	176	−75%

[a] This figure is based on a survey by Perkins in 1988 (Perkins 1991) and Tideman (1987). It was not possible to use the 1991 Census of Population in the area because it was difficult to identify ranch populations within enumeration districts.
[b] This figure is based on the 1991 Census of Population which identified the population of individual ranches. This figure is likely to include, however, 'overspill' population from the service centre of Metsimentle.
Sources: Perkins 1991; Tideman 1987; Questionnaire Survey 1994.

service centres. Table 6.2 reveals population change for the majority of ranches in Study Areas 1 and 2 for which comparable data were available; no previous studies had taken place in Area 3.

There is evidence to suggest that these movements have been subject to different stimuli. Leaseholders exercising exclusive rights to the ranchland forced movement off ranches in Study Area 1, a situation reinforced by ethnic tensions in the area arising from historical and current power relations between ethnic groups. In the late eighteenth century Tswana cattleherders appropriated the springs in the Kalahari and forced the Basarwa into servitude. Many Basarwa became *malata* (slaves) of the Tswana. This system of Basarwa servitude lay at the roots of the development of a successful Tswana pastoral economy allowing masters to build up large herds. In Study Area 2 the persistence of drought affected ranch-owners and dwellers alike as drought-induced death of cattle and coincidental borehole failure rendered the ranches uneconomic. Many of those left on the ranches in this area were women and the elderly who would find it difficult to secure employment elsewhere. The TGLP has increased mobility, producing a population of transients rather than settlers, involved in frequent localized residential moves between the ranches, service centres, and cattleposts. Significantly, one-third of all moves to the ranch areas were from the service centres, which had been created to accommodate ranch populations but have not been able to support their livelihoods.

There is a growing literature that examines the contemporary significance of labour migration (usually rural to urban) and the role of remittance payments in household (predominantly female-headed) livelihoods (see for example Nelson 1992; Kerven 1979; Peters 1983). The TGLP ranch populations represent an interesting deviation from the norm for a number of reasons. Due to the male-oriented nature of ranch work it is mainly men who are mobile, leaving their families behind. There was no evidence, however, that these men were supporting their families through remittance payments due to the lack of waged labour on the ranches, nor was there any evidence of remittance flows to ranch households. Moreover the mobility patterns of respondents revealed there to be no evidence for migration between TGLP ranches and urban areas suggesting that the former may not be integrated within the national migration system characterized by large-scale rural–urban migration.

Implications for Fertility and Population Growth

With development initiatives (e.g. education, schooling, availability of family planning) provided under TGLP there *should* be a shift in intergenerational wealth flows (see Caldwell 1982; Caldwell *et al.* 1987) and fertility *should* decline as children are no longer valued for their labour. The research revealed that where livelihoods are diminished through environmental and social change, rations issued by the government drought relief programme to infants were being appropriated for the whole family's use. Despite the introduction of family planning programmes, children were valued as a source of food supply reducing the incentive to control fertility as the government had replaced the parent as the main provider. This important finding contradicts previous research on drought in Botswana (Rutenburg and Diamond 1993) which suggests drought-associated hardship is conducive to fertility decline.

Family planning and reproductive health initiatives have accompanied the introduction of TGLP service centres. While use of modern contraceptives at 17 per cent, 37 per cent, and 45 per cent respectively in Study Areas 1, 2, and 3 were higher in two of the study areas than the national average of 29 per cent (Botswana CSO 1992), there is evidence to suggest that sterility associated with STD's may be playing an important role in inhibiting population growth. Though mobility is high among male ranch-dwellers, sedentarization at ranches and service centres, even for a few years, was associated by both respondents and health practitioners with increased promiscuity and a high incidence of STDs resulting in the termination of childbearing among sexually active non-contracepting women midway through their reproductive lives. These findings, which require further investigation, have potentially wider implications for future rural African population growth scenarios.

The Relationship between Environmental Opportunities and Health Status

Numerous examples can be gleaned from the literature on how sedentarization often has negative impacts on the health of a population. Concentration of settlement around boreholes or other new sites may, for instance, heighten exposure to certain diseases (Hildebrand 1985). For example, Foster and Anderson (1978: 16) point out that 'The health of hunter-gatherer peoples is . . . affected positively by their nomadic habits; a people few in number and constantly on the go is less likely to reinfect itself from its faecal and other matters than is a large settled population where, once infection is endemic, it is almost impossible to eradicate it short of the most modern environmental sanitation.' A further study by Kent and Dunn (1996) in the Kutse area demonstrated how morbidity has increased with sedentarism/aggregation. They provide compelling evidence that at the newly sedentary community of Kutse the deterioration of health and increase in the prevalence of disease is due to the change from a nomadic to a sedentary lifestyle. On the TGLP ranches and within service centres, health officials reported a high incidence of STDs and HIV which they attributed to the common practice of taking multiple sexual partners and a reluctance to use condoms. Residents were reluctant to discuss these issues and to attend clinics, preferring to consult traditonal doctors in the first instance.

Hunting and gathering activities were considered important in all three ranch areas, yet edible resources have decreased in each, due to both the establishment of ranches and drought. The imposition of hunting restrictions has further compounded the situation, reducing the protein content of ranch residents' diets. As a result, the chief health issues were hunger-related and were attributed by respondents to diminishing entitlements. For example, on the Ncojane ranch block prolonged drought had rendered gathering activities untenable as the testimony of a Basarwa woman living on one of the ranches illustrates (Box 6.1). Many respondents were unaware of the

---

Box 6.1. Gathering for food: The impact of TGLP and drought in the Ncojane ranch block

---

'Sometimes I go gathering. I collect *moretlwa* and *motsotsojane* during the rainy season. When the season is dry like this one I never go out to gather. I only gather after the rains. These wild plants that I have mentioned are few in number these days. There are no longer a lot of them on the ranch, but there is more *moretlwa* than *motsotsojane*. Even after the rains you find that there is more *moretlwa* than *motsotsojane*. But nowadays there is no *moretlwa*. This is a bad year. The plants are no longer around because there has been no rain. The drought is very bad this year. These days we are dying. We suffer from starvation and thirst. . . . When I was young I stayed with my mother. I survived by gathering *motshia*, *mokwa*, and *moretlwa*. I ate these plants. I also dug up *mahupu* (large wild mushrooms). I survived by eating these things I have mentioned. But now I don't gather these things. There is no *mokwa* or *motshia*. Even when I go around the other ranches I never find them. Our lives in the past were better than the life I am leading now. I used to gather wild plants and cook them on top of the hot coals. These days I'm suffering because I can't find these wild foods.' **(Female Basarwa resident in her fifties)**

provisioning under the TGLP of clinics and healthposts, which although important, are not sufficient on their own to ensure the health and well-being of ranch populations as the pills and medicines they dispense cannot cure the nutritional deficiencies that form the underlying cause of so many of their health problems. For example, an elderly Basarwa man on one of the eastern Kalahari ranches explained:

Our main health problem is hunger. People here are starving. Can you not see that? I wake up every day feeling hungry and my whole body aches. There is nothing I can do. I must just get up and go and see the cattle. The mobile clinic comes here but what can they do? They don't give us food. They only come here to weigh the children. Sometimes they bring food rations for the small children but not for others . . . The only medicine for hunger is food.

Livelihoods under the TGLP in Relation to Environmental Opportunities and Entitlements

A summary of contemporary livelihood activities under TGLP is provided in Table 6.3. Despite the commercialization of livestock production, the research revealed a low level of involvement in waged labour on the ranches. Low wages and irregular payment resulted in household incomes well below the national average (Fig. 6.2).

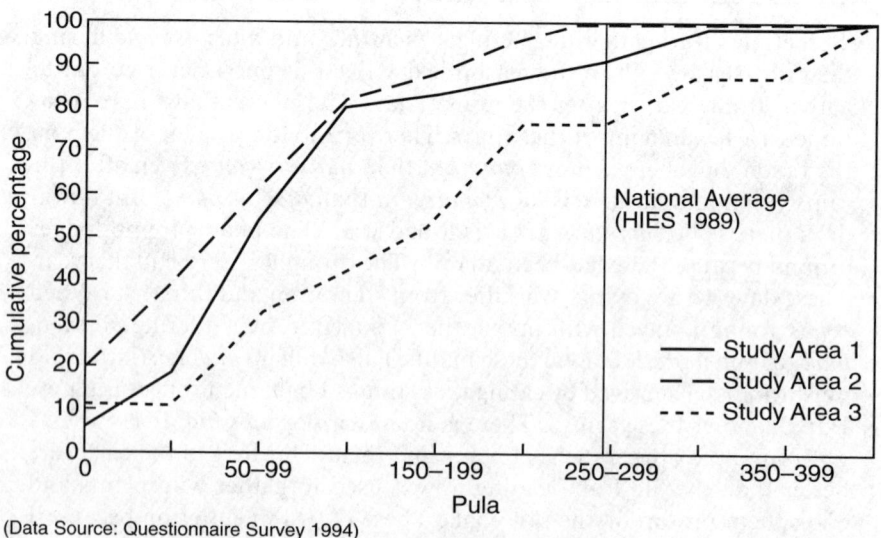

(Data Source: Questionnaire Survey 1994)

Fig. 6.2. Total monthly household income (from all sources).

Table 6.3. Indicators of contemporary livelihood activities under TGLP

	Adults who gather (M/F%) and no. of species collected	Frequency and location of gathering	Involvement in paid labour on ranch	Household ownership of livestock/ smallstock (% who own at least 1 animal)	Food rations from ranch-owner. Type and % who receive regular monthly rations
Area 1	59% (68%/49%) 1. Food = 16 2. Firewood = 11 3. Medicine = 2	Daily on ranch	49% paid 7% unpaid 44% no work	60% cattle 68% donkeys 1% horses 18% goats	Maizemeal 32%
Area 2	25% (34%/19%) 1. Food = 8 2. Firewood = 5 3. Medicine = 1 4. Sale = 4	Weekly on ranch	28% paid 23% unpaid 49% no work	38% cattle 42% donkeys 38% horses 27% goats	Maizemeal 43%
Area 3	—[a]	—[a]	41% paid 27% unpaid 32% no work	55% cattle 22% donkeys 33% horses 22% goats	Maizemeal 78%

[a] Study Area 3 is excluded as only one person gathered.
Source: Questionnaire Survey and Case Studies.

In the absence of cash incomes, ranch populations attempted traditional livelihood practices but found they were unsustainable. Only a handful of respondents in each of the areas admitted to hunting because of fear of persecution as a result of strict new hunting laws, and a decline in wildlife was widely reported. Wildlife decline in the Kalahari has been widely associated with TGLP and the establishment of veterinary fences that cut across seasonal wildlife migration routes (see Ch. 4; Owens and Owens 1980; Williamson and Williamson 1985), yet among rural populations, for whom the hunting of wildlife once represented an important livelihood activity, decline is linked to drought and variability within the physical environment. Almost 40 per cent of residents in the eastern Kalahari identified a change in wildlife in the area although the majority reported the decrease in their hunting activities, as part of their livelihoods, was due not to wildlife declines but to the introduction of hunting permits and to the fencing of ranches. The fencing of TGLP ranches excluded game from the area and limited the range of hunting trips which in the past involved forays lasting several days.

The environmental survey revealed that veld products gathered for human consumption, notably berries and cucurbits, appeared abundant on many ranches in Study Area 1, but were conspicuously absent in Study Area 2 and limited in occurrence in the most arid location, Study Area 3. Despite the relative abundance of veld products in Study Area 1, many residents still reported difficulties gathering, and in some cases were banned from doing so by the lessee. In Study Areas 2 and 3, the lower level of gathering reflects in part the drought-induced depletion of species which potentially poses a serious threat to livelihoods in areas where average cash wages are lower (Fig. 6.2, Table 6.3). Case studies revealed, however, that the provision of food support by ranch lessees and organized 'traditional' systems of mutual support and exchange among residents served as a buffer in times of livelihood stress. In Study Area 1, where environmental resource endowments were greatest during the study period, livelihoods were more stressed as household support systems were observed to be breaking down, as groups are brought together in work-based rather than kin-based communities. Relations with ranch-owners were found, in many cases, to be strained and as a result less assistance in the form of food rations was provided. These findings support previous work on Basarwa groups in Study Area 1 (Hitchcock *et al.* 1993).

The Resourcing of Dependency

The Oxford Food Studies Group (1990: 12) states that welfare-dependency is manifest in under-resourced areas with inadequate opportunities and is a situation that prevails even in non-drought times (see also Hitchcock *et al.* 1993). Welfare support from ranch-owners, and increasingly the government, was found to play a crucial role in determining the livelihood security in TGLP ranch areas. Many of the 'transients' who had moved to and from Service Centres did so in the hope that they could participate in labour-based drought relief programmes—the only real source of employment in these settlements. Similarly infant food rations are increasingly shared between family members who have no other means of support.

The Response of People to Environmental and Land-Use Changes

The ecological surveys (Thomas and Sporton 1997; Thomas *et al.* 2000) show that during the study period the plant resource base was significantly different in the three areas, demonstrated by Detrended Correspondence Analysis and Canonical Correspondence Analysis (Fig. 6.3). On individual ranches, piospheric vegetation zonation (Perkins and Thomas 1993*a*) was marked in both Study Areas 1 and 2 but greatest in the former. Study Area 1 also possessed the greatest edible plant resource opportunities during the investigation. These factors may be due to the location of areas along the climatic

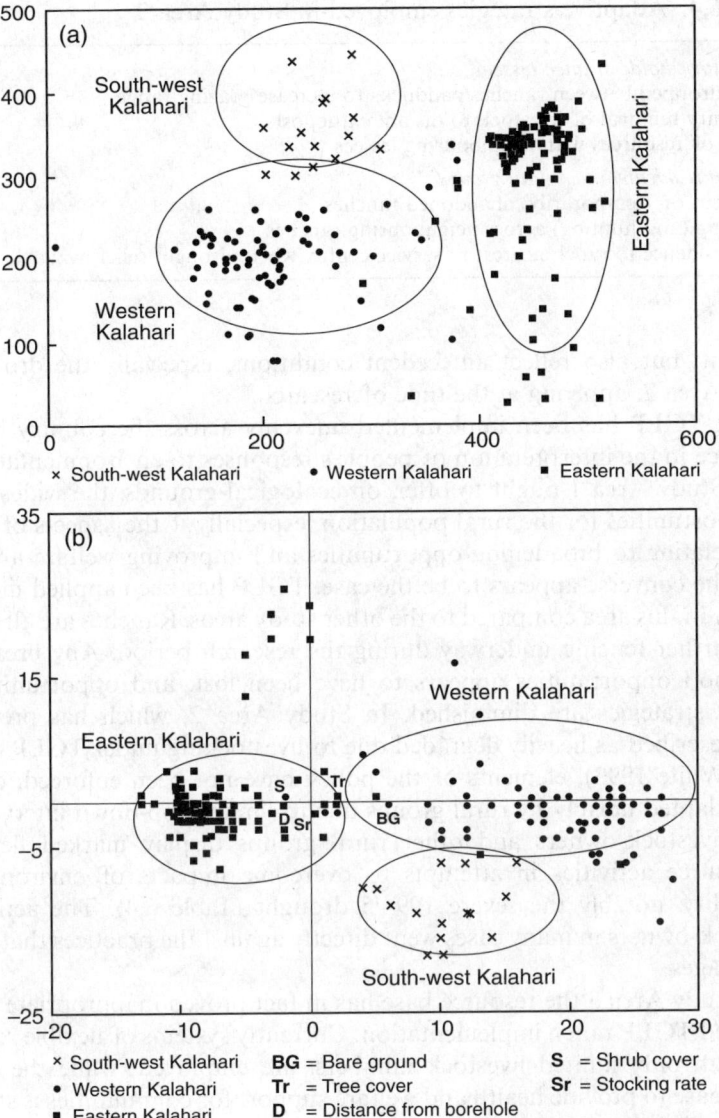

Fig. 6.3. (a) DCA for Kalahari ecological data, giving good separation of sample locations from each study area; (b) DCCA identification of main criteria contributing to the separation of the three study areas (after Thomas *et al.* 2000).

Table 6.4. Adaptive strategies employed in Study Area 2

By livestock holders/ranch lessees
Fences dropped between ranches/paddocks to increase grazing range
Temporary removal of livestock to distant cattleposts
Pooling of resources with neighbouring lessees

By ranch residents/other rural groups
Settlement on (temporarily) abandoned ranches
Gathering (and hunting) across neighbouring ranches
Move residence to ranch nearest to service centre, where drought relief available

---

gradient, but also reflect antecedent conditions, especially the drought in Study Area 2, applying at the time of research.

That TGLP has been implemented unevenly across the country has significance in the interpretation of people's responses to environmental conditions. Study Area 1 ought to offer, on ecological grounds, the widest range of opportunities for the rural population, especially if the aspects of TGLP aims relating to broadening opportunities and improving welfare are to be met. The converse appears to be the case. TGLP has been applied most rigorously in this area compared to the other study areas. Ranches are all fenced, with further fencing underway during the research period. Any breadth to livelihood opportunities appears to have been lost, and opportunities for flexible strategies are diminished. In Study Area 2, which has previously been described as heavily degraded due to livestocking under TGLP (Cooke 1985; White 1993), elements of the policy have not been enforced, or have been adapted flexibly by rural groups due to limited top-down intervention. Both livestock-owners and other rural groups display marked flexibility in resource activities in attempts to overcome impacts of environmental variability, notably the severe 1994/5 drought (Table 6.4). The actions of livestock-owners in many cases went directly against the practices that TGLP encourages.

In Study Area 3 the resource base has in fact proved inappropriate for the intended TGLP ranch implementation. Currently systems of flexible resource use, with only limited livestock numbers, are employed, while the service centre base to provide health and welfare support for communities is stronger than in the other two areas.

The Nature of Poverty and Dependency in the Kalahari

Botswana has emerged as one of the dominant economic powers in sub-Saharan Africa and while per capita income levels remain high, they mask extreme inequalities in wealth between rich and poor (Good 1992). Poverty is a culturally defined multifaceted concept which must be understood within

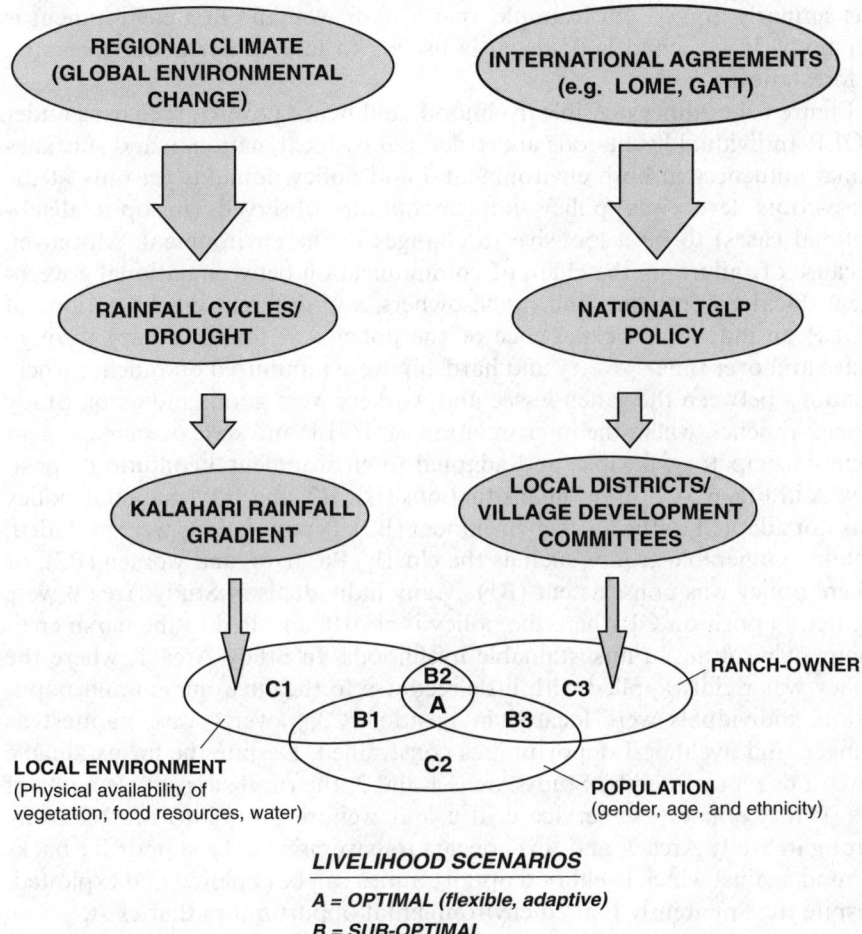

Fig. 6.4. Optimizing livelihoods under conditions of environmental and structural land-use change.

specific social, cultural, and environmental contexts yet all too often external assessments, based for example on per capita income, ignore the *reality* of what it is to be poor. In TGLP ranch areas, respondents associated poverty with the failure to meet their most basic needs such as food, shelter, health, and clothing. At this most basic level when people's coping strategies operate from one day to the next, it is naïve of policy-makers to assume that ranch populations will be concerned with a longer-term vision involving sustainable management of resources. Antrobus (quoted in Braidotti *et al.* 1994: 119) has similarly argued, for example, that a poor woman's first environment is her body. If her child is dying, it is useless to talk to her about trees and deforestation.

Figure 6.4 outlines possible livelihood, and hence poverty, scenarios under TGLP. Individual livelihoods are structured by local, national, and supranational influences in both environmental and policy domains yet only at the grass-roots level was policy implementation observed (in optimal/sub-optimal cases) to be responsive to changes in the environment. Moreover, because of failures in the chain of communication between national government, local government, and ranch-owners, and different interpretations of TGLP, an individual's experience of the policy was found to vary through space and over time. Poverty and hardship were minimized on ranches where relations between the ranch lessee and workers were good, and as on Study Area 2 ranches, where the interpretation of TGLP and style of management were participatory, flexible, and adapted to environmental conditions (position A in Fig. 6.5). Sub-optimal situations (B1, B2, and B3) resulted if policy was not adapted to the local environment (B3), if populations were excluded, notably vulnerable groups such as the elderly, Basarwa, and women (B2); or where policy was non-existent (B1). Many individuals in Study Area 3 were located in position C1 where the policy is clearly unsuited to the harsh environment resulting in unsustainable livelihoods. In Study Area 1, where the policy was rigidly applied with little recourse to the environment and population, individuals were located in position C3, poverty was manifest as hunger, and livelihood opportunities constrained. Despite the unsustainable situations represented by Study Areas 1 and 3, the rural support elements of the TGLP policy (e.g. service centre and welfare provision) are relatively strong in Study Area 3, and this appears to have provided a supportive background against which livelihood opportunities can be explored and exploited, despite the apparently limited environmental opportunities that exist.

How May Populations Respond to Future Change?

Overall, findings suggest that a single policy, which has gained significant notoriety in terms of its predicted negative societal and environmental impacts (e.g. Hitchcock 1978; Cooke 1985), has been subject to marked spatial variability in implementation and impacts, and in one area the aims have been subverted and new resource-use–environment relationships have been manipulated by local populations (Thomas and Sporton 1997). Communities and households continue to practise adaptive traditional activities *within* the framework of policies that elsewhere delimit activities more rigorously. This begins to suggest the factors that may be important in determining population responses to future climatic changes, which may themselves narrow the base of environmental opportunities across the region. What, for example, would be the responses of rural people to greater occurrences of drought and lower overall effective rainfall in, say, Study Area 1? Under the current situation, the opportunities for residents to respond to drought appear to be severely limited, as ranch management is directed towards a fairly inflexible livestock-dominated strategy. Flexible approaches operating in Study Area 2 suggest a regime that allows both livestock activities and other rural practices to operate to some degree. However, in terms of TGLP livestock system goals, Study Area 2 is a failure and Study Area 1 a relative success, since the former operates traditional strategies akin to the cattleposts that TGLP, and more recently the 1991 National Policy on Agricultural Development, seek to replace with more 'modern' approaches. If, as climate models predict, soil moisture levels and rainfall decrease over the coming decades, TGLP approaches to resource utilization may prove impractical and impossible to operate. If this is the case, conditions may evolve (or return), where flexible strategies are the only routes for rural livelihoods in all sectors of society. The 1994 UN Convention to Combat Desertification encourages recognition of traditional local knowledge and approaches to resource use in the context of limiting environmental degradation and reducing poverty. Degradation has not been directly considered in this project, but ultimately the convention may provide a framework in which coping, livelihood, and resource-use strategies that benefit rural peoples *and* the environment gain predominance.

Acknowledgements

The research described in this chapter was funded by the Economic and Social Research Council under Phase III of the Global Environmental Change Programme (Ref. L320253174) and was supported by a research permit from the Office of the President, Government of Botswana. The support of others associated with the project, notably Dr Jean Morrison, is acknowledged.

References

BEHNKE, R. H., SCOONES, I., and KERVEN, C. (1993), *Range Ecology at Disequilibrium* (London: ODI).

BLAIKIE, P., and BROOKFIELD, H. (1987), *Land Degradation and Society* (London: Routledge).

BRAIDOTTI, R., CHARKIEWICZ, HAUSLER, S., and WIERINGA, S. (1993), *Women, The Environment and Sustainable Development* (London: ZED).

BRAAT, L. C., and OPSCHOOR, J. B. (1990), 'Risk in the Bostwana range-cattle system', in J. A. Dixon, D. E. James, and P. B. Sherman (eds.), *Dryland Management: Economic Case Studies* (London: Earthscan), 153–74.

CALDWELL, J. C. (1982), *Theory of Fertility Decline* (London: Academic Press).

CALDWELL, J. C., and CALDWELL, P. (1987), 'The cultural context of high fertility in sub-Saharan Africa', *Population and Development Review*, 13: 409–37.

CAMPBELL, A. C. (1986), 'The use of wild food plants and drought in Botswana', *Journal of Arid Environments*, 11: 81–91.

CAMPBELL, A., MAIN, M., and ASSOCIATES (1991), *Western Sandveld Remote Area Dwellers* (Gaborone: Noraid).

COOKE, H. J. (1985), 'The Kalahari today: A case of conflict over resourse use', *Geographical Journal*, 151: 75–85.

CROLL, E., and PARKIN, D. (eds.) (1992), *Bush Base: Forest Farm. Culture, Environment and Development* (London: Routledge).

CSO (1992), *Health Statistics, 1986–1991* (Gaborone: CSO).

DEBENHAM, F. (1952), 'The Kalahari today', *Geographical Journal*, 118: 12–23.

DOUGILL, A. J., HEATHWAITE, A. L., and THOMAS, D. S. G. (1999), 'Soil water movement and nutrient cycling in semi-arid rangeland: Vegetation change and system resilience', *Hydrological Processes*, 12: 443–59.

FOSTER, G. M., and ANDERSON, B. G. (1978), *Medical Anthropology* (New York: John Wiley).

GANDY, M. (1996), 'Crumbling land: The postmodernity debate and the analysis of environmental problems', *Progress in Human Geography*, 20: 23–40.

GOOD, B. (1992), 'Interpreting the exceptionality of Botswana', *Journal of Modern African Studies*, 30: 69–95.

HILDEBRAND, D. (1985), 'Child mortality and the care of children in rural Mali', in A. Hill (ed.), *Population, Health and Nutrition in the Sahel: Issues in the Welfare of West African Communities* (London: Kegan Paul).

HITCHCOCK, R. K. (1978), *Kalahari Cattle Posts: A Regional Study of Hunter-Gatherers, Pastoralists, and Agriculturalists in the Western Sandveld Region, Botswana* (Gaborone: Government Printer).

——(1985), 'Water, land and the evolution of tenure and administrative patterns in the grazing areas of Botswana', in L. Picard (ed.), *The Evolution of Modern Botswana* (Nebraska: Rex Collins/University of Nebraska Press), 84–121.

HITCHCOCK, R. K., EBERT, J. I., and MORAN, G. (1993), 'Settlement patterns, of the Bakgalakgadi', paper presented at the Symposium on Settlement, Botswana Society, Gaborone.

KENT, S., and DUNN, D. (1996), 'Anaemia and the transition of nomadic hunter-gatherers to a sedentary lifestyle: Follow-up study of a Kalahari community', *American Journal of Physical Anthropology*, 93/3: 455–72.

KERVEN, C. (1979), *Urban and Rural Female Headed Households' Dependence on Agriculture* (Gaborone: CSO).

KILLICK, T. (1991), 'Notes on Macroeconomic Adjustment and the Environment', in J. T. Winpenny (ed.), *Development Research: The Environmental Challenge* (London: ODI).

——(1995), 'Flexibility and economic progress', *World Development*, 23: 721–34.

LEACH, M., and MEARNS, R. (1991), *Poverty and Environment in Developing Countries. An Overview Study, Final Report to the ESRC and ODA* (Brighton: IDS).

——(1996), 'Environmental change and policy: Challenging received wisdom in Africa', in M. Leach and R. Mearns (eds.), *The Lie of the Land: Challenging Received Wisdom on the African Environment* (Oxford: James Currey), 1–33.

McGOWAN INTERNATIONAL, and COOPERS LYBRAND (1988), 'National Land Management and Livestock Project: Incentives/Disincentives Study', 1–3 (Gaborone: Report prepared for the Ministry of Agriculture).

MEARNS, R. (1991), 'Environmental implications of structural adjustment: Reflections on scientific method', *IDS Discussion Paper*, 284.

MOGWE, A. (1992), 'Who was (t)here first? An assessment of the human rights situation of Basarwa in selected communities in the Gantsi District, Botswana', *Occasional Paper*, 10 (Gaborone: Botswana Christian Council).

MOYO, S., O'KEEFE, P., and SILL, M. (1993), *The Southern African Environment* (London: Earthscan).

NELSON, N. (1992), 'The women who have left and those who have stayed behind: Rural–urban migration in Western and Central Kenya', in S. Chant (ed.), *Gender and Migration in Developing Countries* (London: Belhaven).

OWENS, M., and OWENS, D. (1980), 'The fences of death', *African Wildlife*, 34: 25–7.

OXFORD FOOD STUDIES GROUP (1990), *Report on the Evaluation of the Drought Relief and Recovery Programme, 1982–90* (Gaborone: CSO), i.

PEET, R., and WATTS, M. (eds.) (1996), *Liberation Ecologies. Environment, Development, Social Movements* (London: Routledge).

PERKINS, J. S. (1991), 'The Impact of Borehole Dependent Cattle Grazing on the Environment and Society of the Eastern Kalahari Sandveld, Central District, Botswana', Ph.D. thesis, University of Sheffield.

PERKINS J. S., and THOMAS D. S. G. (1993a), 'Spreading deserts or spatially confined environmental impacts? Land degradation and cattle ranching in the Kalahari Desert of Botswana', *Land Degradation and Rehabilitation*, 4: 179–94.

——(1993b), 'Environmental responses and sensitivity to permanent cattle ranching in semi-arid western central Botswana', in D. S. G. Thomas and R. J. Allison (eds.), *Landscape Sensitivity* (Chichester: Wiley), 273–86.

PETERS, P. (1983), 'Gender, development cycles, and historical process: a critique of research on women in Botswana', *Journal of Southern African Studies*, 10: 100–22.

——(1994), *Dividing the Commons: Politics, Policy and Culture in Botswana* (Charlottesville: University Press of Virginia).

REDCLIFT, M. (1995), 'The environment and structural adjustment—lessons for policy interventions in the 1990s', *Journal of Environmental Management*, 44: 55–68.

REPUBLIC OF BOTSWANA (1991), *National Policy on Agricultural Development: Government Paper No. 1 of 1991* (Gaborone: Government Printer).

RUTENBURG, N., and DIAMOND, I. (1993), 'Fertility in Botswana: Recent decline and future prospects', *Demography*, 30: 143–59.

SANDFORD, S. (1980), 'Keeping an eye on TGLP', *NIR Working Paper*, 31 (Gaborone: University of Botswana).

SCOONES, I. (1992), 'Land degradation and livestock production in Zimbabwe's Communal Areas', *Land Degradation and Rehabilitation*, 3: 99–114.

THOMAS, D. S. G., and SHAW, P. A. (1991), *The Kalahari Environment* (Cambridge: Cambridge University Press).

THOMAS, D. S. G., and SPORTON, D. (1997), 'Understanding the dynamics of social and environmental variability: The impacts of structural land use change on the environment and peoples of the Kalahari, Botswana', *Journal of Applied Geography*, 17: 11–27.

THOMAS, D. S. G., SPORTON, D., and PERKINS, J. S. (2000), 'The environmental impact of livestock ranches in the Kalahari, Botswana: Natural resource use, ecological change and human response in a dynamic dryland system', *Land Degradation and Development*, 11: 327–41.

TIDEMAN, K. (1987), *A Report on the Population Survey Carried Out in the Area Designated for the Third Phase Development of TGLP Ranches in Sandveld Central District* (Serowe: District Officer of Lands).

TIFFEN, M., MORTIMORE, M., and GICHUKI, F. (1994), *More People, Less Erosion: Environmental Recovery in Kenya* (Chichester: John Wiley).

TSIMAKO, N. (1991), *Tribal Grazing Land Policy (TGLP) Ranches. Performance to Date* (Gaborone: Ministry of Agriculture).

TWYMAN, C., SPORTON, D., and MORRISON, J. (1999), 'The final fifth: Autobiography, reflexivity and interpretation in cross-cultural research', *Area*, 31/4: 313–25.

WALKER, B. H. (1985), 'Structure and function of savannas: An overview', in J. C. Tothill and J. J. Mott (eds.), *Management of the World's Savannas* (Canberra: Australian Academy of Science).

WHITE, R. (1993), *Livestock Development and Pastoral Production on Communal Rangeland in Botswana* (Gaborone: Botswana Society).

WILLIAMS, M. A. J., and BALLING, R. (1995), *Interactions of Desertification and Climate* (London: Edward Arnold).

WILLIAMSON, D., and WILLIAMSON, J. (1985), 'Botswana's fences and the depletion of Kalahari wildlife', *Parks*, 10: 5–7.

7

Entitled to a Living:
Opportunity and Diversity in the
Wildlife Management Areas

Chasca Twyman

Introduction

This chapter will examine sustainable livelihoods within the context of Kalahari Wildlife Management Areas (WMAs). WMAs are multiple-use areas combining wildlife conservation with the creation of economic opportunities, jobs, and incomes for rural populations (DWNP 1999). Within WMAs rural populations are permitted access to a range of wildlife and natural resources (land, plants, water, and wildlife). WMAs are just one part of Botswana's development strategy to combine sustainable management of environmental resources with the development of the local and national economies. This chapter will explore the extent to which the Kalahari WMAs are changing the access to, and effective use and management of, the natural resources of the rural populations living within these areas. These changes have important implications for the dynamics of livelihood strategies and thus the viability of resource-based livelihoods.

Conceptual debates surrounding livelihoods and livelihood analysis now recognize the diversity of activities in which people are involved (Toulmin 1991; Chambers 1995, 1997; Adams and Mortimore 1997). The livelihood opportunities open to people, and the diverse portfolios of activities that make up 'a living', are now key areas of conceptual and empirical research that are increasingly taking a more central place in policy-making. Such approaches are inherently responsive to people's own interpretations of, and priorities for, their livelihoods, indicating a shift towards a more integrated context-specific grass-roots approach (Carney 1998). They specifically highlight working *with* people and their existing strengths and constraints rather than adopting prescribed donor-driven solutions to poverty and sustainable livelihood issues.

This convergence of policy and livelihoods has been widely recognized and signifies a new era of policy-making that aims to be 'inclusive', 'participatory', and more 'appropriate' to the needs of different people (Forsyth and Leach 1998; Tsing *et al.* 1999; Agrawal and Gibson 1999; Leach *et al.* 1999). This multifaceted approach to policy-making has been particularly noticeable in relation to natural resource management. Within Botswana this shift is encapsulated in the creation of WMAs and the subsequent plans for community-based natural resource management projects. Community-based natural resource management initiatives are currently in vogue around the world. They follow a wider move within development discourse away from prescriptive top-down policies and practices towards grass-roots community development and local natural resource management, thus they have much in common with the recent work on sustainable livelihoods.

The research findings presented in this chapter stem from research between 1994 and 1997. A mixed-methods approach was adopted drawing on a range of disciplines, approaches, and techniques. Oral testimonies provided the main source of data: informal, semi-structured, and repeat interviews, group discussions, and informal conversations. These were complemented by observations, participation on trips (e.g. gathering of wild foods), and the use of secondary sources. Consideration was given to issues of positionality and reflexivity in the cross-cultural research process and the problems and possibilities of interpretation and translations. These are explored more fully in Twyman *et al.* (1999). Lists of species used by local populations were compiled in accordance with the Economic Botany Data Collection Standard (Cook 1995) and qualitative assessments of vegetation and rangeland condition were made in and around the settlements (Twyman 1997; cf. Milton *et al.* 1998). Fieldwork was conducted over a two-year period incorporating a range of seasons as well as a 'good' and 'bad' year in terms of rainfall.

This chapter will first explore more fully the role of WMAs in Botswana's conservation and development strategy. From this context the chapter will focus more closely on the Kalahari WMAs of Ghanzi District in western Botswana, a comparatively resource-poor region of Botswana. This discussion will provide the policy context and background for the main body of the chapter that examines the livelihood dynamics of the rural populations living in the Okwa WMA in western Botswana. The livelihood opportunities and diversity of activities are revealed through case studies of rural households living in two remote settlements in the heart of the WMA.

WMAs in Botswana

Linking community development to wildlife management is increasingly being seen as the way forward both for the establishment of self-sustaining economies in the remote areas and to fulfil the objectives of the wildlife conservationists. Countries such as Zimbabwe have approached this through their community development project CAMPFIRE (see Child and Peterson 1991; Martin 1994*a*, 1994*b*; Zimbabwe Trust 1991), while other countries such as Namibia and South Africa are establishing similar initiatives (Cumming 1990; May 1998). In an attempt to bring conservation and development together the government of Botswana proposed that 20 per cent of the land in the country should be zoned for this dual purpose. Thus in 1986 WMAs were established so that Botswana now has a natural resource management system comprising national parks, game reserves, forest reserves, and WMAs (Fig. 7.1).

WMAs emerged in the 1980s as a result of a national land assessment and zoning exercise. Some of the problems associated with the Tribal Grazing Land Policy (see Ch. 6) led to the reserve category of land to be dispensed with and WMAs to be introduced (Hitchcock 1999). They were designated as areas in which natural resource use (both consumptive and non-consumptive) would be the primary economic activity. Regulations for land and resource use were developed and existing settlements and livestock grazing was accommodated in the WMAs, in consultation with the appropriate local authorities.

Since wildlife is a state resource in Botswana, citizens may only hunt if they have licences obtained from the Department of Wildlife and National Parks, the government body with overall responsibility for wildlife resources. Portions of the country are divided into a number of Controlled Hunting Areas (CHAs), which are designated for a variety of uses: community-based, commercial safari, and photographic safari amongst others. As commercial hunting and photographic safaris are carried out almost entirely by private companies (many with headquarters outside the country) a number of CHAs were designated for community-controlled hunting to promote the participation of local people in wildlife management and tourism (Hitchcock 1999). Many, but not all, of these community-controlled hunting areas fall within the boundaries of WMAs.

Local communities, NGOs, and development agencies have already begun planning and implementing a number of projects in some of the community-controlled hunting areas. Some communities have obtained wildlife quotas from DWNP after forming a Quota Management Committee. Communities are then able to either utilize the quota directly themselves, or lease all or part of it to a safari company (for consumptive or non-consumptive uses). If the latter, i.e. a joint venture, then the community must establish a legal entity such as a trust or company for their group. They must also display representativeness and accountability to DWNP. Once communities have gained proprietorship over the wildlife resources in this way, they can establish enterprises that can provide benefits such as employment and markets for their products, thus enhancing their livelihood options (ibid.).

This move to involve local people more directly with the management of

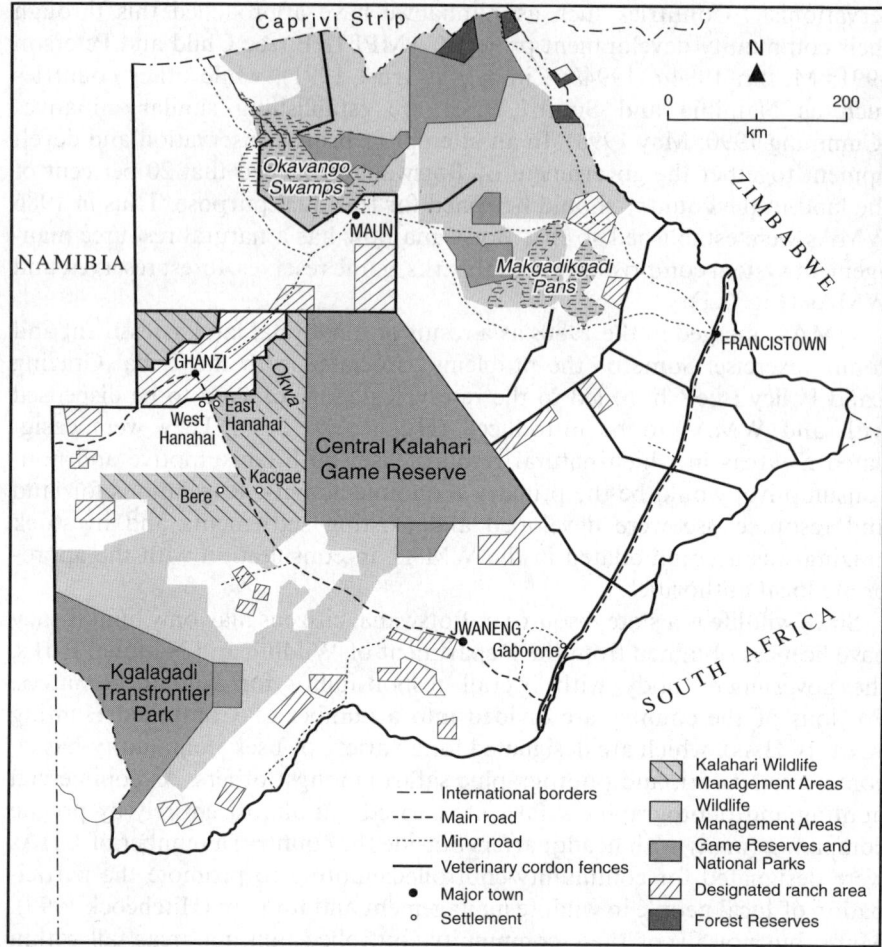

Fig. 7.1. Wildlife Management Areas and Protected Areas in Botswana.

wildlife and other natural resources is being crystallized by the proposed community-based natural resource management policy. Drafted in 1999, the government highlighted a 'policy gap': existing natural resource policies in Botswana did not define the objectives of community-based natural resource management or provide firm guidance for its implementation, despite it being the guiding principle for many of the recent government initiatives (NRMP 1999). However, the government recognizes the 'vital importance of conservation strategies that are national and ecosystem in perspective and yet local in approach' (ibid.: 9), and the draft policy lays out a set of objectives and guidelines for community-based natural resource management.

WMAs still remain central to these shifts in policy, as they are the primary areas for wildlife utilization. Resources within the WMAs can vary enormously depending on the characteristics of the region. Generally WMAs in the northern areas near the Okavango Delta are considered resource-rich with significant economic potential through tourism. Areas further west within the drier Kalahari environment are considered resource-poor in comparison and have less potential for major tourist developments. These Kalahari WMAs form the focus of the next section.

Wildlife Management Areas in the Kalahari

Ghanzi District lies in the west of Botswana. Figure 7.2 identifies the current land-use zones in the district as well as the location of the WMAs and settlements in the district. Two-thirds of the district (67%) is zoned for wildlife conservation or management (DLUPU 1995). Ghanzi District is the most remote district in Botswana and poor infrastructure has perpetuated its image in the past as isolated and backward. Now, linked to the rest of Botswana by the Trans-Kalahari Highway, economic opportunities in the town and district are expanding. It has a diverse population comprising Batswana, Bakgalagadi, Basarwa, European, and Afrikaner groups and an economy based on the cattle industry (Molamu 1987; Adams *et al.* 1989). The vegetation

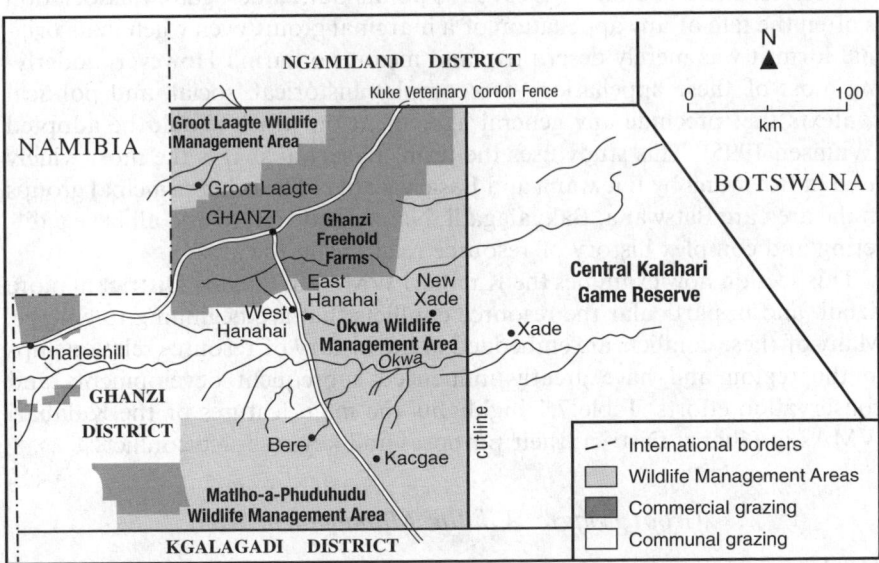

Fig. 7.2. Land-use zones and settlements in Ghanzi District.

of the area is classified as northern tree and bush savanna (and central Kalahari bush savanna in the south of the district) according the scheme established by Weare and Yalala (1971) and has low species diversity. Across part
of the north-east of the district lies a limestone ridge, the Ghanzi ridge, which
is the centre for the commercial cattle industry in this region. The surrounding area has low relief except where incised with fossil river valleys. The water
table is more accessible along these drainage lines and this has influenced settlement distribution. This is further enhanced by plant and vegetation preferences for the valley.

This area of Botswana has a complex history of overlapping and competing claims for environmental resources. This history has transformed the
resource relationships in the region influencing people's contemporary interactions with the environment and its resources. It also underpins the emergence of unequal access to and effective use of resources necessary for the
livelihoods of the indigenous rural populations. The people residing in
western Botswana come from a range of different ethnic groups. Within the
field-study area the majority of the rural population are Basarwa, also known
as 'the first people' or 'aboriginal people'. The Basarwa of Botswana have
also variously been known as Bushmen, San, or Khwe, and Remote Area
Dwellers (RADs) (Mogwe 1992; Good 1993; Campbell and Main 1991;
Hitchcock 1996), and there are a number of subgroups (see Ch. 1, Fig. 1.1)
each with a separate language (Naro, G/wi, !Xo, Kaukau, and G//ana are
the principal subgroups encountered in this research). These terms all have
different connotations to both the users of the terms and to the people
themselves, and as Sanders (1989: 174) points out, 'a derogatory association
is often the fate of any appellation of a marginal group, even when in its original form it was merely descriptive and meant no harm'. However, underlying most of these appellations are complex historical, social, and political
contexts that preclude any general agreement on which should be adopted
(Wilmsen 1995). This study uses the term 'Basarwa' as it is the most widely
used in Botswana by Batswana and Basarwa alike. The other principal groups
in the area are Batswana, Bakgalagadi, Nama, and Herero and all have a differing and complex history of resource relations in the area.

This section now examines the Kalahari WMAs in Ghanzi District in more
detail, and in particular the resource conflicts that are beginning to emerge.
Many of these conflicts are embedded in the history of resource relationships
in the region and have greatly influenced subsequent developments and
conservation efforts. Table 7.1 highlights the main features of the Kalahari
WMAs in Ghanzi District, their purposes and key resource conflicts.

Groot Laagte Wildlife Management Area

The Groot Laagte WMA contains two settlements. Groot Laagte itself is
situated 87 km north-east of Ghanzi. The population comprises mainly Naro

Table 7.1. The Kalahari Wildlife Management Areas

	Groot Laagte	Matlho-a-Phuduhudu	Okwa
Location	North-west Ghanzi District. Enclosed by Ghanzi freehold farms to south-east, Kuke veterinary fence to the north, and fenced Namibian border to the west	Southern Ghanzi District. Borders the communal grazing area to the north, Ghanzi–Lobatse road to the east, and the Kgalagadi District boundary to the south. Unfenced	East Ghanzi District. Adjacent to the Central Kalahari Game Reserve in the east and the Ghanzi–Lobatse road in the west. Northern boundary with the fenced Ghanzi freehold farms and unfenced Kgalagadi District boundary to the south
Size	3,908 km^2	8,816 km^2	13,613 km^2
Resident population (estimated)	<900	<400	<1,000
User population (estimated)	>1,300	>700	>2,000 (includes residents of West Hanahai)
Official settlements	Groot Laagte, Qabo	Bere	East Hanahai (north), Kacgae (south); New Xade (east)
Unofficial settlements	—	Matlo-a-Phuduhudu Takatswane	Zero, Okwa, Lone Tree, Lokalane, Rooibak
Main ethnic groups	Kaukau, Naro, Bakgalagadi	!Xo, Bakgalagadi, G/wi	Naro and G/wi (north); G/wi, !Xo, Bakgalagadi (south)
WMA purpose	Wildlife management and utilization; disease buffer zone	Wildlife management and utilization; wildlife migration corridor from south-west Botswana	Wildlife management and utilization; wildlife migration corridor from south-west Botswana; buffer zone for Central Kalahari Game reserve
Main resource conflicts	Encroachment from Ghanzi freehold farms by cattle; Qabo situated in area known for *Dichapetalum cymosum* which is poisonous to cattle	Cattle trek route from Ncojane TGLP farms runs through the centre of the Wildlife Management Area leading to considerable encroachment from the cattle on veld resources	Encroachment in the north from Ghanzi freehold farms, and in the southern area from Qushe and Morgan boreholes. Conflict in south with proposed Botswana Defence Force training area within the WMA. Increased pressure on resources with the establishment of New Xade settlement

Source: Republic of Botswana 1992; Environmental Services 1992; RPM 1995*a, b*.

and Kaukau Basarwa groups with some Bakgalagadi having moved in more recently. Qabo is one of the more recent (August 1995) settlements to be established under the local government Land and Water Development Scheme for the Ghanzi Farm Basarwa (1976 LG 32 (v)). The establishment of Qabo was initially controversial with land-use conflicts coming high on the local political agenda when non-Remote Area Dwellers moved into the settlements with large numbers of livestock.

The livelihood base for the residents in this WMA is unique. Both Okwa and Matlho-a-Phuduhudu WMAs are predominantly unfenced, except where they border the Ghanzi Freehold Farms or the Ncojane Tribal Grazing Land Policy ranches. This allows for free movement of wildlife and in particular the preservation of the wildebeest migration from the Schwelle in the Matlho-a-Phuduhudu WMA and Kgalagadi District up to the Central Kalahari Game Reserve through the Okwa WMA (Crowe 1995). However, Groot Laagte is completely fenced, limiting the resources available in the area and making it particularly susceptible to encroachment from the surrounding farms. Encroachment by cattle in the area alters the wild foods available and changes the grazing material available for wild animals. This is particularly important around Groot Laagte itself where harvesting of *morama* (a prolific and commercially valuable nutritious veld food) is the most significant local commercial activity.

Qabo highlights several land-use issues of relevance to all the Ghanzi WMAs. First, the complexity of settlement rights in Remote Area Dweller settlements is illustrated in Qabo through the differing interpretations of these rights by Basarwa, Bakgalagadi, and government officers. Second, the problems of cattle-grazing and water rights in and around these settlements is exacerbated in Qabo by the ambiguities of the settlement rights. Third, dispossession of resources is occurring with 'outsiders' coming in with cattle businesses and buying up cattle, or simply moving in with their own cattle. This alters the resource base of the area and therefore has an impact upon the livelihoods of other people living in the settlements. In Qabo, people from Ghanzi and those on nearby farms, both of whom had depleted their own range resources, moved into the settlement within weeks of its establishment. The Principal Game Warden of Ghanzi District commented, in 1995: 'these people are now over grazing the ranches and moving out into the WMAs. For example, the case of Qabo and the people from Kuke and D'kar. Others have taken many cattle to this place. It is not just cattle, goats are a big problem too in this area.'

The local government took action and within six months the large livestock owners had returned with their animals to their farms. The conflict was recognized by the government principally because of the potential severity of the situation. The Groot Laagte WMA was seen as a disease buffer between Ngamiland and the Ghanzi Freehold Farm block. Uncontrolled livestock-rearing around Qabo and illegal livestock movement would increase the likelihood of diseases such as Bovine Pleuro-Pneumonia spreading into Ghanzi District and jeopardizing the country's main source of regular beef production. Without this threat it is unlikely that the government would have acted so swiftly. The 'Qabo issue' is very similar to the cattle problem in Kacgae in the Okwa WMA but in this latter case no firm action is being taken.

Such conflicts illustrate the diversity of issues enabling and constraining the rural population living off the natural resource base. The WMAs are intended partly to safeguard the interest and rights of non-stock-owners affected by the Tribal Grazing Land Policy (Republic of Botswana 1986). Allowing such conflicts to continue unresolved therefore jeopardizes these intentions, reflects a lack of commitment to the policy, and perpetuates the conflicts between RADs and the larger cattle-owners.

In an effort to gain greater control over their area (and their natural resources) the residents of Groot Laagte have chosen to take part in the government's community-based natural resource management programme. The residents understand they can apply for rights over Community-controlled Hunting Areas (CHAs) and they have been successful. With the help of the Kuru Development Trust (an indigenous people's NGO), the communities of Groot Laagte and Qabo formed the Huiku Trust in May 1999 and are currently in the process of deciding which natural resource management options would best benefit their community and surrounding area.

Matlho-a-Phuduhudu Wildlife Management Area

Bere is the only official RAD settlement in this WMA. It was initially set up under private auspices and later brought under the Land and Water Development Scheme for the Ghanzi Farm Basarwa (1976 LG 32 (v)). Initial developments in the settlement were heavily influenced by an ethnobotanist to the overall detriment of the settlement (Biesele *et al.* 1989). Local people were not able to sustain the projects without assistance and thus failed to establish a means of making a living. Thus Bere has a history of failed projects and initiatives which has left the population sceptical of new projects and policy changes. They depend heavily on their natural resource base but face encroachment from the trek route that runs through the Wildlife Management Area as well as from the adjacent communal grazing areas. Despite evidence that wildlife numbers are low in the WMA (RPM 1995*b*) hunting remains an important occupation. Sales of crafts also provide a significant income (Gjern 1994). As with Kacgae in the Okwa WMA, the apparent lack of resources does not seem to hinder people's reliance on the resource base. With few alternatives it makes the necessity of living off the land more acute. This WMA also provides an important link between the Schwelle in the south-west and the grazing ranges and water of the northern central Kalahari in the dry and drought seasons (Verlinden and Masunga 1993; Verlinden 1993).

Okwa Wildlife Management Area

The Okwa WMA is the largest of the Ghanzi WMAs and was gazetted in 1992. It contains three official settlements, East Hanahai and Kacgae, managed under the Land and Water Development Scheme for the Ghanzi Farm Basarwa (1976 LG 32(v)), and New Xade, recently established after residents were relocated outside the Central Kalahari Game Reserve. There are also several informal settlements that have arisen through the sedentarization of people around government boreholes (e.g. Rooibak, Lone Tree, Okwa, Lokalane, and Zero). Kacgae was established in 1973 after a growing number of !Xo and G/wi Basarwa had become resident around government boreholes (Lone Tree and Lokalane) and was later brought under the government scheme. East Hanahai was established directly under the scheme in 1982 principally for the Naro Basarwa from the north-eastern Ghanzi Freehold Farms near D'kar. Many of the residents had been initially residing in West Hanahai for several years awaiting the completion of East Hanahai. New Xade was established in 1997 after the main body of this research was conducted. However its location in the Okwa WMA is significant and should be considered in all discussions concerning the control and sustainable use of natural resource in the region.

Rooibak, in the north of the WMA, was a potential settlement site under the local government's Land and Water Development Scheme for the Ghanzi Farm Basarwa (1976 LG 32 (v)). Rooibak was identified by Ghanzi Farm Basarwa and was informally settled before drilling had been completed. The borehole was found to be low yielding and the water unpotable, such that the residents were asked to suggest an alternative area. A new site, Qabo, with potable water was found in the Groot Laagte WMA. However, some people at Rooibak refused to move to the new site as they did not consider it 'their land'. They continue to remain at Rooibak illegally hunting and gathering and keeping a few small livestock. The land around them is subject to severe encroachment from the surrounding farms, further constraining the livelihood options available. Nearby freehold farmers are reported to 'drop their fences' and let their cattle graze in the WMA, watering them from the edge of their farms. There is also evidence to suggest that fences are dropped and wild animals pushed into the farms. The extent of this illegal use of resources is currently unknown but is severe enough to effect the resources available for subsistence hunting and gathering in the valley. The government will not recognize the settlement and therefore provides no support such as hunting licences or transport assistance for schoolchildren. The Dutch Reformed Church at D'kar provides fuel for the borehole engine and people have to travel over 60 km by poor sand track to D'kar for services such as the school and clinic.

New Xade was established in 1997 amid international media attention. The relocation poses major risks for people in the Okwa WMA (Hitchcock, pers. comm. 1999). The population size is substantial, some 700 people. The quality of the land and the density of the resources in the new settlement are seen as insufficient to sustain the people who were brought to the new location. There are few useful trees and shrubs in the vicinity of the new settlement forcing people to travel extremely long distances in order to obtain fuel and other critical resources. Another problem is water, which is not available locally; instead, it has to be piped from a borehole some 60 km away.

The Okwa WMA forms the focus for the remainder of this chapter. Case studies of respondents from East Hanahai and Kacgae are examined in detail below, exploring the links between livelihood opportunities, resource availability, and the diversity of practices.

Livelihood Opportunities and Diversity: Cases from Okwa

Framed by the conceptual debates surrounding livelihoods (illustrated in Ch. 1), and the policy context outlined above (and in Ch. 3), this section examines resource-based livelihoods in a Kalahari WMA in western Botswana. In the Kalahari context, empirical research has shown that themes of place and identity are clearly important in how people perceive resources and thus how this affects their relationship with the environment and their 'ability' to follow certain livelihood pathways. Issues of inequality, dependency, and ethnic tension also signify the complexity of these resource relationships and underpin resource conflicts evident here, and around the world (Moore 1993, 1996; Young 1995; Silvern, 1995; Huggins *et al.* 1995). Focusing on livelihoods, and being responsive to people's own interpretations and priorities, empowers the various levels of enablement and constraint in resource utilization for livelihoods to be distinguished and thus potential effects of community-based natural resource management scenarios identified. Livelihood strategies can then be explored to determine their relative importance to different people and to understand their dynamics with critical implications for proposed policy changes. In particular this will show how people manage to make a living in an environment of opportunity, diversity, conflict, and co-operation, and individual case studies are used to illustrate the dynamics of the resource networks involved in the livelihood options. Common themes between individuals, groups, and settlements emerge as well as distinct differences and disparities. These accounts draw on field research carried out in Kacgae and East Hanahai between 1995 and 1997.

Cattle

Rearing of cattle is universally acclaimed by respondents to be the most important livelihood opportunity in the settlement. All respondents aspire to own more cattle and see this as a way out of poverty or, what they term 'suffering'. Cattle are seen as problem-solvers as their sale or slaughter can provide cash at short notice and can ease hunger in difficult times: 'Because when you have an animal if you have some problems you can take a calf and sell it to buy maize meal for the children. Sometimes if you are hungry you can slaughter it and share it with the children' (interview with Morris, 1995). Thus cattle are resources that can provide subsistence food income, cash income, and assets. In East Hanahai livestock provide a good livelihood opportunity and do 'work' for many respondents, though there are recognized difficulties. Moses' case study gives an insight into a successful livelihood strategy which is based primarily on cattle (Box 7.1). The emphasis is placed on cattle though the dynamics of, and interactions with, other livelihood options is clearly visible. Cattle in his case are viewed as long-term assets and regular income, freeing labour commitments (e.g. not being dependent on waged work) and thus enabling Moses to invest in alternative livelihoods such as *motoroko* gathering in the Ghanzi Freehold Farms. Moses is unique among the people interviewed in this research in that he established and maintained contacts in the Freehold Farms and this enabled him to pursue specific activities unavailable to others in East Hanahai. His livelihood strategies are the mainstay of the household. His wife was terminally ill at the

---

Box 7.1. Moses' livelihood profile, East Hanahai

Moses is Herero and is about 60 years old. He has lived in East Hanahai since it was established and was formerly living and working in the Ghanzi Freehold Farms. He was born in Namibia, spent some time working in the mines near Johannesburg but returned to Botswana to settle. His main livelihoods are cattle rearing, hunting, cultivation, and a *motoroko* plot.

Moses has an established set of livelihoods that enable him to survive well in times of drought and other stresses. He has kept up his network of contacts in the Ghanzi Freehold Farms maintaining relationships with farm-owners ensuring him access to grazing. His livelihood of cattle means that he does not have to participate in waged labour and thus has time to invest in his *motoroko* plot. He is keen to invest his time here and acknowledges the five years before he will be able to harvest cochineal beetles. A further incentive is provided by food handouts associated with maintaining plots in the scheme. Despite being Herero by origin, he associated himself closely with the Basarwa, particularly in terms of land dispossession. Moses has a secure livelihood portfolio and is able to experiment with planting wild foods. However, his wife was ill in 1995 and he could not leave the settlement for any length of time, and thus hunting was not an option for him.

beginning of the research and thus was unable to contribute to subsistence needs: she died in mid-1996. His children, all adult, live in the settlement, married locally, and have permanent waged employment. Moses' secure livelihood position benefits those around him, enabling them to obtain skilled wage employment in the settlement and to accumulate cattle.

In Kacgae the rearing of livestock is rarely viable as respondents lack the capability to exploit this livelihood option. They are constrained by their lack of command over their relationship with the Bakgalagadi, while the Bakgalagadi see the Basarwa as a profitable market to exploit. Though they are able to sell their cattle to pay debts, by doing this they are denying themselves their entitlements to keep cattle as long-term assets. This is precisely the scenario in which Tsharu (the second case study) found himself. Tsharu is G/wi and has lived in the Kacgae area throughout his life. Allocated five cattle in 1992 under the livestock scheme, he quickly 'lost' them all through straying and in payment of debts. In some ways it could be argued that Tsharu is too poor to own cattle: the overriding daily necessities of food overwhelm desires to retain cattle as long-term assets. Box 7.2 illustrates the limited set of opportunities, in comparison with Moses in East Hanahai, that Tsharu is able to exploit.

Both case-study respondents face considerable encroachment of cattle into and around their settlements. People from Ghanzi graze cattle in both areas. When summoned to retrieve their cattle, these people claim that the cattle have gone astray accidentally. Other government workers living in the settlements keep cattle but rarely admit to keeping many.

Box 7.2. Tsharu's livelihood profile, Kacgae

Tsharu is aged about 40 years old. He is G/wi and is married with several children. He has lived in and around Kacgae all his life. His main livelihoods are drought relief labouring, gathering, limited hunting, and sporadic cultivation.

Tsharu has been allocated cattle through the government scheme but has not been able to keep them, he has sold them all. Because of the limited number of places available on the drought relief scheme he can work only every other month. Thus he tries to gather wild foods to supplement their living. He and his wife rely heavily on the monthly food handouts provided by the 'Under 5s' scheme, but this is not enough for two adults and six children. He hunts occasionally with relatives but not regularly because he has no horse or dog. When rains are good Tsharu is able to harvest small amounts of maize, beans, and sorghum. In years of poor rainfall, there are few safety nets for Tsharu's household.

These people who came here through the government, who are supposed to be the ones to develop this place, they are the ones who buy the things from the people, the teachers, the policemen, drivers because they are the ones who get a lot of money. All those people are having their cattle here and councillors also. They keep their cattle here on the place which is supposed to be for the Basarwa. What the Basarwa get from the government is almost nothing because what they get is only the school to keep their children and water only. There is nothing else. There are no toilets in the village and there is also no grass, everything is dead. The environment has been captured by these people. The people who are brought here to keep us.   (interview with Judas, 1995)

Judas, a young Mokoba man from Kacgae, explains his interpretation of this situation. Once more he promotes a view that is sympathetic to the Basarwa who are present in the discussion. It is interesting that he identifies himself with the Basarwa in the last phrase: 'The environment has been captured by these people. The people who are brought here to keep us.' It is a strong image, revealing the idea that the settlements are places to manage people rather than resources. The following two comments relate to Kacgae and East Hanahai respectively.

Let me say something there. People who are staying in Morgan came from different areas. They come and keep their cattle there because we have clinics this side. They came this side in other words we are almost together. Even if they are there, they still come here.   (interview with Cukuri, 1995)

It's something which is *hanahasgosie* (a common thing), these people who come from elsewhere to let their cattle graze here. Right now we are crowded. Now because the *matimelas* (stray cattle project) has started maybe now they will get decreased. They will become afraid and keep their cattle. With grass and grazing it's a wide thing (tcha = wide, i.e. important).   (interview with Cixa, 1995)

This last phrase, 'with grass and grazing it is a wide thing', highlights a serious problem in both settlements. In East Hanahai grazing is considered a problem and cattle are trekked several kilometres from the settlement to find grass. Trees and shrubs around the settlement are heavily grazed and the settlements themselves are bare, mirroring the encroachment areas and sacrifice zone observed around cattleposts (Perkins and Thomas 1993). Grazing and water rights are 'free' to Remote Area Dwellers living in these settlements yet they have little effective command over them. The resources are not in the control of residents in these settlements and thus they do not have the power to *exclude* resource users nor *regulate* resource use. While they may have *apparent* power to regulate and exclude they do not have the *actual* power to do so. This is the arena in which resource conflict occurs. Bakgalagadi, Herero, and Batswana cattle-owners who graze their cattle illegally in East Hanahai are exploiting this opportunity. By doing this they are distorting the entitlements of the Remote Area Dwellers. Moses' case illustrates this well. He keeps relatively few cattle in the settlement, all of which were originally bred from cattle obtained through the livestock scheme. However he also keeps cattle on a friend's cattlepost in the communal grazing area and the local government are encouraging him to apply for his own borehole to accommodate his cattle. Though he owns a relatively large number of cattle in total (over thirty), Moses is still classified as a Remote Area Dweller in East Hanahai and thus receives a hunting permit and was able to receive destitute relief for his wife when she was alive.

The large numbers of cattle in the settlement use valuable grazing and water resources, altering other people's abilities to make effective use of these resources. In Kacgae this situation is more acute. Here, though rights to resources are available, paradoxically the ability to make use of this endowment has been removed. A history of conflict and subordination of the Basarwa by Bakgalagadi has shaped their images of each other and subsequent interactions. Thus Bakgalagadi view Basarwa as a market to exploit, and the Basarwa absorb this exploitation with apathy, easily done in a time of complex renegotiation of relationships during a time of sedentarization and change. This is essentially a dereliction of understood behaviour. Though the sentiment of regulation is there, the capability of the residents to enforce restrictions and exclusion from access to resources is weak.

Gathering

The gathering and use of plants and tree products is based on complex sets of rights and relationships in the settlements, rather than simply on the physical supply of resources (Twyman 1997). Traditional gathering activities among the Basarwa populations have been well documented (Lee and DeVore 1976; Heinz and Maguire n.d.) with a more recent focus on changing social practices linked with the sedentarization of formerly mobile groups (Hitchcock and Marks 1991; Guenther 1996; Kent 1996; Vierich and Hitchcock 1996). Though Guenther (1976) has highlighted the notion of nostalgia and the revitalization of the past, this body of literature has not addressed the incongruence that emerges from reconstructions of past gathering practices as central to present-day Basarwa identity.

Looking more closely at how respondents view gathering in the Okwa WMA, there seems to be some inconsistency, an incongruence between past and present. Gathering is regarded as highly important within the overall strategy for making a living. However, in practice it rarely amounts to the degree of significance that has been implied, even accounting for the seasonality of the fieldwork. This suggests that gathering is of symbolic as well as practical importance, thus redefining the interpretation of availability, rights, and use of these resources. These two levels of importance are not explicitly distinguished, hence the incongruity, and emphasis is continually placed on past resources and gathering experiences, i.e. the symbolic significance it represents in present-day life.

The following quotations by Tcibi, a Naro woman from East Hanahai, also imply a remembered past of plentiful resources in comparison with the present-day 'sufferings'.

They are scarce nowadays. Long back it was better but now they are very scarce. There are not even *khuts'uu* this side.   (1996)

In this place we live on things like *nxam*. This one was there in the older days and now the only one we are living on is *caa*. The *caa* is the one which we live on in this place. Have you not written it down because I showed it to you once. *Nxam* is not around this place.   (1995)

Thus gathering has importance to many people, identifying them with their past and giving value to traditional activities. The two case studies (Boxes 7.3 and 7.4) illustrate these respondents' deep and strong identification with gathering and their considerable expertise and knowledge about plants and foods. This is emphasized through their repetition of past stories about gathering trips. However in both cases, after good rainfall, neither respondent opts for the long gathering trips they talked about at length in the preceding dry period. 'I didn't go myself I just saw some of the people bringing *mahupu* and other things. They gave them to me. The !Xo also went on their side and collected some wildfoods. And the G/wikhwe also collected some wild foods' (interview with Soana, 1996).

Both these respondents opt for livelihoods that they see as more important in terms of options for making a living, thus marginalizing gathering to the periphery to be fitted in around the labour demands and requirements of

---

Box 7.3. Tcibi's livelihood profile, East Hanahai

Tcibi is a widow in her late fifties/early sixties. She is Naro. Her husband died several years ago and she remained in East Hanahai with her sisters and children. She was born in the Ghanzi Freehold Farms, worked as a maid for several years, married and moved to Hanahai in the early 1980s. Her main livelihoods are gathering, sporadic cultivation, tending a *motoroko* plot, drought relief labouring, crafts, and livestock (one cow).

Tcibi has a limited set of livelihood opportunities, especially if compared with other more wealthy residents. Though she has no access to male labour for hunting or the collection of ostrich eggshells she does have access to the labour of her sisters and daughters. This means she (and her relatives) are comparatively prolific cultivators when rainfall permits and several of them have *motoroko* plots. Tcibi reported in 1995 that if the rains fell in 1996 she would return to the bush and live off the wildfoods. However, after good rains in 1996 she reported that though she managed to collect wildfoods nearby, she was unable to go on the long trips planned as she had her farm to tend and her drought relief work. She even delayed her trip to the *morama* belt (15–20 km north of East Hanahai) in order to finish harvesting and processing her melon crop.

Box 7.4. Soana's livelihood profile, Kacgae

Soana is !Xo and is in her mid-forties. She was married but became a widow in 1996. She has always lived in the Kacgae area and was one of the first people to settle at the new borehole in the 1970s. Her main livelihood activities are gathering, work as a school cook, drought relief labouring.

In Kacgae, Soana illustrates a more radical shift in livelihood options during the fieldwork period. In 1995 Soana lived with her husband Tsao in a small traditional grass hut with their two youngest children. At this time they survived on his destitute relief rations with Soana supplementing this with a few gathered wild foods (though very few were available) and begging within the settlement. She said that if the rains were good she would return with her family to the bush for several months and eat plentifully. Soana was unable to get a place on the drought relief project despite the monthly rotation of employment. In early 1996 Tsao died and she moved out of their hut. Within a week she had managed to secure a job as a school cook and a month later obtained a place on the drought relief project, labouring. She now had no time for gathering and was not even able to build a new hut. She lived under a tree near the edge of the settlement. Gathering for Soana in 1995 had been viewed as the most important thing that could relieve her family's hunger. By 1996, with the introduction of waged work, gathering was viewed as something others had time for, marginalized in place of waged work.

other livelihoods. While this has been observed in other studies (e.g. Hitchcock and Marks 1991; Kent 1996) the critical issue here that has rarely been addressed is the central, but incongruent, role gathering still plays in present-day Basarwa identity. There is an incongruence of how values for livelihoods are ascribed and how these views ultimately shape resource relationships.

The lifestyle of sedentarization has shifted the labour patterns of gathering (Barnard 1996a, b). Small bands are no longer on the move and their 'usable', 'accessible' environment has essentially reduced (Solway and Lee 1990). Boundaries shift as people follow livelihoods (such as waged labour or cultivation) which tie them to specific places and thus reduce the length of time they can be away from the settlement and the distance they can travel. Lifestyle changes such as dependency on fresh water, requirements for children to attend school, pay-days for waged work, and distribution days for clinic food handouts or *motoroko* food handouts all require people to remain in close proximity to the settlement (Hitchcock and Holm 1993; Hitchcock *et al.* 1993). Thus resource boundaries are shifting, conceptual boundaries of the usable accessible environment are changing. Consequently the realized availability of resources is also changing.

The regional, historical, and ethnic significance of resources is in constant flux. Traditional structures are reshaping and reforming, revealing a dissolution of traditional patterns of gathering (cf. Hitchcock 1996; Kent 1996). There is still a 'culture' of gathering and a wealth of knowledge about plants and trees. However the importance of gathering has been distorted by its perceived centrality to people's lives. Gathering is seen as central to livelihood strategies but in practice is used as something to fall back on in times of stress when alternative livelihoods fail. Thus gathering remains central to the construction of people's identities rather than central to the construction of livelihood strategies.

Notions of access to resources are contingent upon how they are viewed, e.g. what is there and how it can be used. The boundaries of what is perceived as available to different people are continually shifting according to other livelihoods that enable or constrain people in particular ways. In parallel to these shifting resource boundaries are shifting boundaries of command and control over resources. People are also aware of regulations that attempt to regulate their use of resources, usually by altering practices or by excluding people from using resources. The following quotation from a group discussion illustrates the severity of the constraint that people feel is controlling their lives: 'If you want to pick berries you have to take a permit, if you want to cut down a tree you have to take a permit, if you want to dig out a plant you have to take a permit but the Lord doesn't say this. It's just the blacks who own the things' (1995).

This quotation illustrates the bureaucratic shadow under which people feel their rights to resources are being controlled. It also introduces the notion that the land and resources are under the control of the government and the Bakgalagadi. Contradictions become apparent in peoples' livelihood strategies and conflicts begin to emerge.

The principal livelihood strategy beginning to emerge is one of diversity and opportunity, making the most of available options when possible rather than relying on one livelihood. This form of diverse and complex strategy is well recognized for resource-poor areas (Chambers 1997), particularly in the Kalahari context (Hitchcock 1996; Solway and Lee 1990). The significance in this context is the implications that diversity and complexity carry for the 'success' of the WMAs.

Hunting

Endowments to hunt animals are complex and comprise both formal and legal rights as well as informal, illegal, yet personally legitimized rights. The legitimization is based on the exigent need to hunt, not only to provide sustenance but also to maintain a connection with tradition and to bring this connection into a present-day context. To make effectual the use of these endowments, to realize the entitlements, requires a multitude of conditions to be available. Labour requirements must be weighed up with other livelihood demands and access to the physical means of hunting must be actualized, e.g. horses, dogs, transport, hunting partners, availability of animals etc. In parallel to this are the motivations of hunting, influencing the viability of hunting as a livelihood. Here, the distorting effects of NGO craft markets are significant, providing a valuable outlet through which cash income can be obtained. However, such markets are subject to change and flux and are not regular and reliable. They are also influenced by government policy, tourist demand, and other macro-economic dimensions.

The disposal of hunting products through sale and sharing networks is under considerable change. These changes effect female-headed households in particular (Hunter *et al.* 1990). Though women are eligible for hunting permits few are issued to them as, according to one respondent, 'women do not hunt'. Problematic within this gendered and permitted hunting system is the collection of ostrich eggshells (though this is currently under review) and this has significant implications for single-headed households (cf. Peters 1986 in relation to Tswana households).

Inherent in many of the comments about hunting are references to the importance of hunting to the construction of Basarwa male identity. Recognizing the danger of assuming a homogeneous group of male Basarwa, it should be noted that these constructions were expressed to different degrees by different people and clearly vary considerably, for example, between older and younger generations and between the two settlements. Also strongly emphasized is the alienation from the mechanisms of control over this resource and thus over the viability of this livelihood as part of an overall subsistence strategy, a recurrent theme through the discussion of livelihood strategies.

The two case studies chosen here illustrate the very different notions of hunting within young men's livelihood strategies. Boxes 7.5 and 7.6

Box 7.5. Morris's livelihood profile, East Hanahai

Morris is Naro, in his twenties, and lives alone. He has lived in East Hanahai all his life and attended primary school in the settlement. His main livelihoods are drought relief labouring, livestock, and piecework.

Morris does not view hunting as a fully viable livelihood option. Animals are regarded as scarce because of the perceived drought and hunting trips are not viewed as worthwhile in most cases. Morris disparages traditional hunting methods; preferring horses and guns, more modern methods; however these require negotiations with Bakgalagadi and a renegotiation of post-hunting practices. Consequently, Morris rarely hunts, saying instead that he has to do his work on the labour-based drought relief project.

Box 7.6. Sixfour's livelihood profile, Kacgae

Sixfour is G/wi. He is 25 years old and married. He was born in Kacgae and lived with his fathers and brother until the death of his father in 1996, when the household split up. His main livelihoods are hunting, gathering, piecework, and crafts. His wife collects wild foods and works on the drought relief programme when she can.

Sixfour's livelihood strategy radically altered during the fieldwork period. In 1995 Sixfour and his brothers were prolific hunters, having in store up to fifty skins at one time. They hunted together (sometimes with friends), using traditional methods such as trapping and spearing, and processed the skins as a group. They lived as one household combining incomes with their wives, children, and elderly father. The wives worked on the labour-based drought relief project and, as their children were under 5 years of age, they were allocated food from the clinic.

During early 1996 Sixfour's father died and the former household dispersed. The main compound was abandoned and each brother established his own separately. Shortly after this Sixfour became ill and was unable to continue hunting with his brothers. As the households were now separate he was excluded from their immediate sharing networks and only received food when he demanded it. His illness meant that he was unable to work and he was thrown off the labour-based drought relief project amid allegations of laziness. This caused tension between Sixfour and the extension workers and thus he was unable to negotiate destitute relief either. His wife maintained her work on the labour-based drought relief project (alternate months) and gathered veld foods whenever possible. From a successful livelihood strategy in 1995, Sixfour's lifestyle changed radically with the death of his father and the dispersal of the former household. With an illness that makes him unable to work, there are few options for Sixfour and his household. His wife must do all the work to meet subsistence needs. Inevitably such circumstances create tensions and Sixfour exaggerates the household's problems with mounting drinking debts at the shebeens.

respectively show the livelihood portfolio of Morris, a young Naro man from East Hanahai, and Sixfour, a young hunter from Kacgae.

Hunting clearly has a varied and complex place in current livelihood strategies. Therefore blanket changes in policy affecting rights and control over wildlife may have far-reaching, deleterious, and perhaps unpredictable effects in these settlements. Hunting relationships need to be clearly understood if new proposals within the WMAs are to promote viable sustainable livelihoods, and these proposals are to be accepted by Kalahari communities.

Cultivation

Cultivation of melons, beans, and sorghum is another livelihood promoted in the settlement by the government to encourage the self-sufficiency of the rural populations. The principal environmental resources required for cultivation are land and water, with other things such as seeds, ploughs, and donkeys as additional necessities. Cultivation is seen as particularly important as it provides an alternative livelihood when the Labour-Based Drought Relief Programme stops. In practice, these livelihoods often overlap, resulting in conflict. For example, following the rains in 1995–6 a reasonably good harvest of melons was achieved by many people in East Hanahai and people continued to participate in the Labour-Based Drought Relief Programme, tending their plots either in the afternoons or at weekends. The cessation of the Programme in July 1996 in Ghanzi District coincided with end of the harvest. Thus two livelihoods ceased between May and July at a time when few alternatives were available. Though melons had been stored and dried these could not provide a complete diet and could not be sold to provide cash income. Winter is a particularly sparse time of year, wild foods are difficult to find and often lacking in moisture. Animals in the Kalahari region are also highly dispersed at this time of year, making successful hunting more difficult.

Cultivation involves decisions about labour allocation and risk assessments that need to be weighed up against other options. Tsharu, a G/wi respondent in Kacgae, sharply brings to a focus the difficult decisions based on risk and uncertainty that people must make.

Just everything is important because a person is supposed to rear livestock. Those things we want them but we can't get them. . . . When there is no rain there is nothing you can do because there will be no water in the fields. We just stay without doing anything. If the rains come then we plough. . . . We live in a difficult way because we go out to look for things like the one I brought to you last time. We live on those until the rain comes. (1995)

Tsharu emphasizes the central and overriding importance of livestock-rearing: a livelihood that is not considered dependent on rain. All the other livelihood options he refers to in connection with rainfall, bringing in the central importance of climatic variability to livelihood strategies.

Sixfour, the hunter, emphasizes this dependence on rainfall again in the following quotations. Respondents have no control over climatic variability, thus they can only plan contingency options for times of no or variable rainfall. 'If the rains don't come, you don't eat anything. It is *leuba*. In Sesarwa it is called *tcankuri*. It means it is *leuba*. We lead a difficult life. We don't have anything to eat like *moretlwa* and *motsotsojane*. We can't plough and we can't find things like boroku. And *kgengwe* also and *caa'* (1995).

Box 7.7 illustrates the case study of Q'uku (a Naro woman) and the resource networks involved in the *motoroko* project in East Hanahai. The key issue here is the distorting effect the NGOs and donor projects have on the perceptions of land rights in the settlement. Though Q'uku has been able to gain access to a *motoroko* plot through a group application for land rights, she still feels anxious and uncertain about the long-term security of her land rights in the area. She takes a considerable risk in investing time and labour in the project (which she does not fully understand), but a stable livelihood strategy enables her to do so. Q'uku and her husband Cixa have successfully reared cattle through the livestock scheme and have purchased other livestock with money saved from labouring on the drought relief project. Cixa hunts when he has time and Q'uku does craftwork to supplement their income. However the long-term viability of this project and this livelihood is un-certain and the monthly food handouts are certainly significant for Q'uku's participation in the project.

Box 7.8 illustrates the case study of Ankie (a young /Xo woman in Kacgae) and rain-fed agriculture. Ankie is one of the few people in Kacgae to have

---

Box 7.7. Q'uku's livelihood profile, East Hanahai

Q'uku is about 45 years old and is Naro. She was born on the Ghanzi Freehold Farms and grew up in Kuke and D'kar before marrying Cixa and moving to East Hanahai in the early 1980s. Their main livelihoods are cattle rearing, a *motoroko* plot, cultivation, gathering, labouring on the drought relief project, and also piecework.

Q'uku and her husband work hard to keep as many livelihood options open as possible. Both work on the drought relief scheme, often avoid-ing place rotations and thus being enabled to save some money. They keep a few cattle and maintain two field plots: one *motoroko* and the other maize, beans, watermelons, and sorghum if there are reasonable rains. Q'uku also gathers wild foods whenever she is out collecting fire-wood, and makes specific trips for *morama* or *mahupu* if resources are abundant. These she is able to sell in Ghanzi or in D'kar when visiting relatives. Cixa will also do piecework in the settlement. Overall, Cixa and Q'uku have managed to diversify their options to the extent that even in drought years they are able to survive without the risk of destitution. They are not wealthy or secure compared to Moses (Box 7.3), but they are considered to be working hard and doing well by others in the settlement.

Box 7.8. Ankie's livelihood profile, Kacgae

Ankie is /Xo and is 23 years old. She lives on her own with two children, though she has extended family living in the settlement nearby. She was born at Lone Tree but moved to Kacgae as a young girl when the settlement was established. Her main livelihoods are: drought relief labouring, cultivation, and livestock-rearing.

Ankie is struggling to make a living and she has few options open to her. As a single parent she has no access to male labour on her farm, and no one contributing to her household income. She also has little access to hunting products for craft production. She works on the drought relief scheme and when good rains permit, she is able to cultivate maize, beans, and watermelons. Ankie relies heavily on her extended family when times are difficult and there are no rains. Without the drought relief scheme she would be totally dependent on them for food other than that provided by the 'Under 5s' scheme.

planted recently though she harvested few crops. As the drift fence for the farm block is not completed Ankie built a makeshift bush fence to keep out the large numbers of cattle grazing in the area. This requires a significant commitment of labour irrespective of ploughing, planting, and harvesting. The racialized access to land and resources around Kacgae adds to the complexity of the resource relationships surrounding this livelihood. Ankie feels strongly that the G/wi should not farm in /Xo territory. Considerable tension exists between the /Xo and the G/wi over this changing access to land and resources.

The crucial dimension in cultivation in terms of these resource relationships is land. Though land rights have been at the centre of most public debates about the Basarwa (Survival International, 1997; *Guardian* 1996; *The Times* 1996; *Mail* and *Guardian* 1996; *Okavango Observer* 1996) it is rarely addressed publicly to the same extent at the government level. Above, reference was made to Q'uku's uncertainty about her land rights on the *motoroko* project. Here it is argued that the *motoroko* project is *not* introducing a new set of land rights into the settlement, it is merely replicating the land rights in existence for rain-fed agriculture, promoted by the government. Further it is argued that this does not address the fundamental inequalities in land allocation that are historically embedded and still hotly contested, as this comment illustrates: 'And also if you have your own land you can be free to do whatever you want on your own land to make a living. But it is difficult in such a situation whereby this place is owned by the government so it's very difficult' (interview with Moses, 1995).

Moses refers to land and ownership of land, an unusual phrase. He also does not consider his farming and *motoroko* plots as significant land rights, or ownership rights, illustrated by their absence in the ensuing discussion in the interview. This suggests there are two disparate classifications of land in this context. There are small plots allowing the carrying out of specific livelihood options such as *motorokos* and rain-fed agriculture or collectively through Permaculture horticulture projects. Though donor agencies (USAID, NORAD) promote these initiatives as breakthroughs in the way Basarwa land rights are viewed, the Basarwa in this context do not see their land issue as resolved. The land, as referred to here by Moses, is a far greater concept and area than a 200 m by 200 m plot near the settlement. Land in this sense signifies a vast resource on which 'to make a living'. Thus in this context the principle constraints are the lack of control over *land* and the consequent lack of control over *livelihoods*.

This section has revealed the relative importance different people place on different livelihood options. Central to strategies based on natural resources are references to waged work through the Labour-Based Drought Relief Programme. Employment on this programme has had a distorting effect on labour allocation which has influenced significantly the effective use of some natural resources. Clearly then livelihoods are not simply linked to resources through availability. Complex interactions have been discussed and illustrated through case studies and this has highlighted the dynamics and dialectical processes that shape these resource relationships (cf. Harvey 1996). The processes mediating environmental entitlements are more complex, dynamic, and diverse, even chaotic. The people living in these settlements are undergoing major transformations in the institutional frameworks that structure their resource relationships, frameworks that are being rapidly dissolved and reworked to cope with the changing situations in which people find themselves. By linking land and livelihoods, and recognizing the severe problems of power and control that people face at this juncture, it can be seen that this forms the most salient linkage through which resource relationships and livelihood dynamics and strategies can be understood.

Concluding Remarks

This chapter has examined the livelihood strategies in the two settlements, teasing out the dynamic interaction between different options and within the policy frameworks structuring the livelihoods themselves. Livelihood dynamics are complex and diverse and dependent on the capabilities of people to make effective use of their entitlements. For example, there are several levels at which resource users can be excluded from utilizing resources for subsistence and small-scale commercial purposes: (a) national—through government permits (e.g. hunting); (b) local—through informal racialized access to land, wildlife, and veld products (e.g. in Kacgae); (c) specific—through an individual's access to particular forms of labour (e.g. male labour from hunting products). People's access to resources necessary to follow certain livelihood pathways are mediated by institutional dimensions (e.g. permits, regulations, informal rules, and accepted traditions), but they are also mediated by processes of identity, ethnicity, and place, not all of which have institutional dimensions. Consequently, livelihoods are valued and ascribed with meanings that reflect more than subsistence needs and can appear incongruent with orthodox notions of livelihood strategies.

To manage the environment according to indigenous past practices requires the necessary degree of control over the resource base as well as the motivation to do so. In both settlements respondents feel they lack sufficient control over their resource base. Continued marginalization in terms of access to and use of resources is exacerbated by Bakgalagadi and Batswana encroachment of the resources surrounding the settlements, and this is felt most acutely in Kacgae. In parallel with this encroachment and dispossession, respondents are continually stigmatized as old-fashioned, backward, and incapable to the extent that young members of the settlement have begun to reject their traditional values in favour of modern Batswana lifestyles. Young respondents reject their ties to the past and to traditions that are seen by others as the essence of Basarwa identities but inconsistencies emphasize the continual struggle being faced in the construction of these new identities. This rejection is painfully obvious to the older members of the settlements. Powerlessness and inequality are acutely felt among these populations as they are gradually *distanced* from their resource base. These issues underpin the ethos of dependency that has evolved in the settlements and the apparent apathy towards self-help projects that government and NGO workers report.

This analysis has shown the complexity and diversity of people's livelihood strategies within this dryland environment. This has important lessons for WMAs and any proposed community-based natural resource management projects that are planned. Implementation of projects will need to recognize this diversity and enhance rather than constrain the opportunities available. There is already evidence emerging that some of the changes proposed in these areas could result in fundamental changes to livelihood strategies, some with negative consequences for household food security (Twyman 1997, 1998; Taylor 1998).

References

ADAMS, W. M., DEVITT, P., GIBBS, D., PURCELL, R., WHITE, R. (1989), *Western Region Study: A Review of the Development Potential of Kgalagadi and Ghanzi Districts*, Report prepared for Ministry of Agriculture, Gaborone (Oxford: Mokoro).

ADAMS, W. M., and MORTIMORE, M. J. (1997), 'Agricultural intensification and flexibility in the Nigerian Sahel', *Geographical Journal*, 163, 150–60.

AGRAWAL, A., and GIBSON, C. (1999), 'Enchantment and disenchantment: The role of community in natural resource conservation', *World Development*, 27/4: 629–49.

BARNARD, A. (1996a), 'Culture and development: The case of the Bushmen. Paper presented at the University of South Africa, Pretoria, 10th August 1995, *Occasional Papers 57: Research and Development in Bushman Communities: Two Lectures* (Edinburgh: Edinburgh University Department of Anthropology).

——(1996b), 'Keynote Address: The State of the Art in Anthropology and Sociology, Basarwa Research Workshop, 24th August 1995', *Occasional Papers 57: Research and Development in Bushman Communities: Two Lectures* (Edinburgh: Edinburgh University Department of Anthropology).

BIESELE, M., GUENTHER, M., HITCHCOCK, R., LEE, R., and MACGREGOR, J. (1989), 'Hunters, clients and squatters: The contemporary socio-economic status of Botswana Basarwa', *African Studies Monographs*, 9/3: 109–51.

CAMPBELL, A., and MAIN, M. (1991), *Western Sandveld Remote Area Dwellers Report. Socio-Economic Survey: Remote Area Development*, submitted to MLGL and NORAD, Gaborone, Botswana.

CARNEY, D. (ed.) (1998), *Sustainable Rural Livelihoods: What Contribution Can We Make?* (London: DFID).

CHAMBERS, R. (1995), 'Poverty and livelihoods: Whose reality counts?' *IDS Discussion Paper*, 347.

——(1997), *Whose Reality Counts? Putting the Last First* (London: Intermediate Technology).

CHILD, B., and PETERSON, J. H. (1991), *CAMPFIRE in Rural Development: The Beitbridge Experience* (Harare: Branch of Terrestrial Ecology and Centre for Applied Social Sciences).

COOK, F. E. M. (1995), *Economic Botany Data Collection Standard*, prepared for the International Working Group on Taxonomic Databases for Plant Scientists, Royal Botanic Gardens, Kew.

CROWE, T. M. (1995), 'Wildlife utilizations: Science vs. ethics and northern vs. southern perspectives', *South African Journal of Science*, 91: 375–6.

CUMMINGS, D. H. M. (1990), *Communal Land Development and Wildlife Utilization: Potential and Options in Northern Namibia* (Harare: Earth Africa).

DLUPU (DISTRICT LAND USE PLANNING UNIT) (1995), *Ghanzi District Land Use Zoning Plan* (Ghanzi District: Land Use Planning Unit).

DWNP (DEPARTMENT OF WILDLIFE and NATIONAL PARKS) (1999), *Community-Based Natural Resource Management Practitioners Guide* (Gaborone: DWNP and Ministry of Commerce and Industry).

ENVIRONMENTAL SERVICES, (1992), *Okwa Wildlife Management Area Land Use Plan Vol. 1. Main Report and Appendices 1–7*, prepared for the Ghanzi District Land Use Planning Unit (Gaborone: Environmental Services Botswana).

FORSYTH, T., and LEACH, M. (1998), *Poverty and Environment: Priorities for Research and Policy: An Overview Study*, prepared for United Nations Development Programme and European Commission (Sussex: Institute of Development Studies).

GJERN, B. (1994), *Gantsi Craft Report of Training Programme for Gantsi Craft Producers* (Ghanzi: Gantsi Craft).

GOOD, K. (1993), 'At the ends of the ladder: Radical inequalities in Botswana', *Journal of Modern African Studies*, 31/2: 203–30.

Guardian (1996), 'Bushmen of the Kalahari appeal for support in ancestral lands fight', 5 April 1996, UK.

GUENTHER, M. (1976), 'From hunters to squatters: Social and cultural change among the Farm San of Ghanzi, Botswana', in R. B. Lee and I. DeVore (eds.), *Kalahari Hunter-Gatherers: Studies of the !Kung San and their Neighbours* (Cambridge, Mass.: Harvard University Press), 120–34.

——(1996), 'Diversity and flexibility: The case of the Bushmen of southern Africa', in S. Kent (ed.), *Cultural Diversity Among Twentieth-century Foragers: The African Perspective* (Cambridge: Cambridge University Press), 65–86.

HARVEY, D. (1996), *Justice, Nature and the Geography of Difference* (Oxford: Blackwell).

HEINZ, H. J., and MAGUIRE, B. (n.d.), 'The Ethno-biology of the !kõ bushmen. Their ethno-botanical knowledge and plant lore', *Occasional Paper* 1 (Gaborone: Botswana Society).

HITCHCOCK, R. K. (1996), *Kalahari Communities: Bushmen and the Politics of the Environment in Southern Africa*, Document no. 79 (Copenhagen: IEGIA).

——(1999), *Community Based Natural Resource Management in Botswana*, report to the First Nations Development Institute, Fredericksburg, Virginia.

HITCHCOCK, R. K., and HOLM, J. D. (1993), 'Bureaucratic domination of hunter-gatherer societies: A study of the San in Botswana', *Development and Change*, 24: 305–38.

HITCHCOCK, R. K., and MARKS, S. A. (1991), *Traditional and Modern Systems of Land Use and Management and User Rights to Natural Resources in Rural Botswana*, report to USAID-funded Natural Resources Management Programme, Gaborone, Botswana.

HITCHCOCK, R. K., EBERT, J. I., and MORGAN, G. (1993), 'Drought, Drought Relief and Dependency Among the Basarwa of Botswana', in R. Huss-Ashmore and S. H. Katz (eds.), *Africa Food Systems in Crisis: Part One: Micro-Perspectives* (London: Gordon & Breach), 303–36

HUGGINS, J., HUGGINS, R., and JACOBS, J. (1995), 'Kooramindanjie: Place and the postcolonial', *History Workshop Journal*, 39: 165–81.

HUNTER, M. J., HITCHCOCK, R. K., and WYCKOFF-BAIRD, B. (1990), 'Women and wildlife in Southern Africa', *Conservation Biology*, 12/90: 1–10.

KENT, S. (1996), 'Cultural diversity among African foragers: Causes and implications', in S. Kent (ed.), *Cultural Diversity Among Twentieth-century Foragers: The African Perspective* (Cambridge: Cambridge University Press), 1–18

LEACH, M., MEARNS, R., and SCOONES, I. (1999), 'Environmental entitlements: Dynamics and constitutions in community-based Natural Resources Management', *World Development*, 27/2: 225–47.

LEE, R. B., and DEVORE, I. (eds.) (1976), *Kalahari Hunter-Gatherers: Studies of the !Kung San and their Neighbours* (Cambridge, Mass.: Harvard University Press).

Mail and Guardian (1996), 'Kalahari Bushmen face removal', 4–11 April 1996, South Africa.

MARTIN, R. B. (1994*a*), *The Influence of Governance on Conservation and Wildlife Utilization*, paper presented at Conference on Conservation Through Sustainable Use of Wildlife, Brisbane, Australia (Harare: Department of National Parks and Wildlife Management).

MARTIN, R. B. (1994*b*), *Alternative Approaches to Sustainable Use: What Does and Doesn't Work*, paper presented at Conference on Conservation Through Sustainable Use of Wildlife, Brisbane, Australia (Harare: Department of National Parks and Wildlife Management).

MAY, J. (1998), *Poverty and Inequality in South Africa*, report prepared for the Office of the Executive Deputy President and the Inter-Ministerial Committee for Poverty and Inequality (Pretoria: Government of South Africa).

MILTON, S. J., DEAN, W. R., and ELLIS, R. (1998), 'Rangeland health assessment: A practical guide for ranchers in arid Karoo shrublands', *Journal of Arid Environments*, 39: 253–65.

MOGWE, A. (1992), 'Who was (t)here first? An assessment of the human rights situation of Basarwa in selected communities in the Gantsi District, Botswana', *Occasional Paper*, 10 (Gaborone: Botswana Christian Council).

MOLAMU, L. (1987), *Ghanzi Baseline Survey* (Gaborone: Applied Research Unit, Ministry of Local Government, Lands and Housing).

MOORE, D. S. (1993), 'Contesting terrain in Zimbabwe's Eastern Highlands: Political ecology, ethnography and peasant resource struggles', *Economic Geography*, 69/4: 380–401.

——(1996), 'Marxism, culture, and political ecology', in R. Peet and M. Watts (eds.), *Liberation Ecologies: Environment, Development, Social Movements* (London: Routledge) 125–47.

NRMP (NATURAL RESOURCE MANAGEMENT PROJECT) (1999), 'A paper on the promotion of natural resource use, conservation and management: community-based natural resource management policy', draft discussion document produced by Botswana's Natural Resource Management Project, Gaborone.

Okavango Observer (1996), 'Spotlight on Botswana: Hardbattle tells UN of forced removal at Xade', 5 April 1996, Botswana.

PERKINS, J. S., and THOMAS, D. S. G. (1993), 'Spreading deserts or spatially confined environmental impacts? Land degradation and cattle ranching in the Kalahari Desert of Botswana', *Land Degradation and Rehabilitation*, 4: 179–94.

PETERS, P. (1986), 'Household management in Botswana: Cattle, crops and wage labour', in J. L. Moock (ed.), *Understanding Africa's Rural Households and Farming Systems* (Boulder: Westview), 133–54.

REPUBLIC OF BOTSWANA (1986), *Wildlife Conservation Policy (A Paper on the Utilization of the Wildlife Resource of Botswana on a Sustainable Basis)*. Government Paper No. 1 of 1986 (Gaborone: Government Printer).

——(1992), *Wildlife Conservation and National Parks Act. Act 28 of 1992* (Gaborone: Government Printer).

RPM (RESOURCE PLANNING MANAGEMENT) (1995*a*), *Ghanzi District Wildlife Management Area Management Plan. Groot Laagte. Draft Final Report. Volumes One, Two and Appendices*, report prepared for Ghanzi District Land Use Planning Unit as part of the USAID Natural Resource Management Project (Gabarone: Resource Planning and Management).

——(1995*b*), *Ghanzi District Wildlife Management Area Management Plan. Matlho-a-Phuduhudu. Draft Final Report. Volumes One, Two and Appendices*, report prepared for Ghanzi District Land Use Planning Unit as part of the USAID Natural Resource Management Project (Gaborone: Resource Planning and Management).

SANDERS, A. J. G. M. (1989), 'The bushmen of Botswana: From desert dwellers to world citizens', *Africa Insight*, 19/3: 174–82.

SILVERN, S. (1995), 'Nature, territory and identity in the Wisconsin Treaty Rights Controversy', *Ecumene*, 2/3: 267–92.

SOLWAY, J. S., and LEE, R. B. (1990), 'Foragers, genuine or spurious: Situating the Kalahari San in history', *Current Anthropology*, 31/2: 109–46.

SURVIVAL INTERNATIONAL (1997), 'Action: Botswana squeezes Kalahari peoples out' (June) *Urgent Action Bulletin*, UK.

TAYLOR, K. (1998), ' "Where seeing is believing": Exploring and reflecting upon the implications for Community Based Natural Resource Management', *Occasional Paper Series*, 72 (Edinburgh: Edinburgh University, Centre for African Studies).

Times, The (1996), 'Bushmen fight to stay on in last Botswana haven: "The lion and me are brothers. I am confused that I should leave this place" ', 5 April 1996, UK.

TOULMIN, C. (1991), 'Natural resource management at the local level: Will this bring food security to the Sahel?' *IDS Bulletin*, 22/3: 22–30.

TSING, A., BROSIUS, J., and ZERNER, C. (1999), 'Assessing Community-Based Natural Resource Management', *Ambio*, 28/2: 197–8.

TWYMAN, C. (1997), 'Community Development and Wildlife Management: Opportunity and Diversity in Kalahari Wildlife Management Areas, Botswana', Ph.D. Thesis, University of Sheffield, Sheffield.

——(1998), 'Rethinking Community Resource Management: Managing Resources or Managing People in Western Botswana?' *Third World Quarterly*, 19/4: 745–70.

TWYMAN, C., SPORTON, D., and MORRISON, J. (1999), 'The final fifth: Autobiography, reflexivity and interpretation in cross-cultural research', *Area*, 31/4: 313–25.

VERLINDEN, A. (1993), *Movement Patterns of Some Important Kalahari Ungulates in the Kalahari of Botswana Caused by Very Low Rainfall in 1991–1992* (Gaborone: Research Division, Monitoring Unit, Department of Wildlife and National Parks).

VERLINDEN, A., and MASUNGA, G. (1993), *An Update on the Status of Some Wildlife Species in Kgalagadi District: movements, biomass and numbers* (Gaborone: Research Division, Monitoring Unit, Department of Wildlife and National Parks).

VIERICH, H., and HITCHCOCK, R. (1996), 'Kua: Farmer/foragers of the eastern Kalahari, Botswana', in S. Kent (ed.), *Cultural Diversity Among Twentieth-Century Foragers. The African Perspective* (Cambridge: Cambridge University Press), 108–24.

WEARE, P. R., and YALALA, A. (1971), 'Provisional vegetation map', *Botswana Notes and Records*, 3: 131–52.

WILMSEN, E. N. (1995), 'Who Were the bushmen? Historical process in the creation of an ethnic construct', in J. Schneider, and R. Rapp (eds.), *Articulating Hidden Histories: Exploring the Influence of Eric R Wolf* (London: University of California Press), 308–23.

YOUNG, E. (1995), *Third World in the First: Development and Indigenous Peoples* (London: Routledge).

ZIMBABWE TRUST (1991), *People Wildlife and Natural Resources: The CAMPFIRE approach to rural development in Zimbabwe* (Harare: Zimbabwe Trust).

8

Coping with Uncertainty: Adaptive Responses to Drought and Livestock Disease in the Northern Kalahari

Robert K. Hitchcock

Introduction

An enduring but in many ways unfortunate image of Africa is that of a mal-nourished child sitting in the sand at the edge of a refugee camp set in a bleak desert landscape. All too often, Africa is characterized as a kind of basket case, a continent beset by drought, famine, disease, and civil conflict (Kaplan 1996). Less attention tends to be paid in the media to African success stories, to the diverse and ingenious strategies used by communities, families, and individuals to cope with uncertainty and reduce risk, and to the innovative programmes employed by African governments and non-government organizations to predict, plan for, manage, and assess the impacts of changes in their socio-economic and environmental situations.

From the perspective of their residents, southern African countries are subjected relatively frequently to problems posed by drought, livestock disease, and other difficulties. For the purposes of this chapter, drought will be defined as a deficiency of precipitation that results in water shortage for some activity (e.g. plant growth) or for some group (e.g. farmers) (Wilhite and Glantz 1987: 13). Drought can also be viewed as a rainfall-induced shortage of some economic good such as grazing for livestock that is brought about by inadequate or badly timed rainfall (Sandford 1979: 34). Severe droughts, such as those of 1973–4 in the Sahel and 1961–5 in Botswana, can result in great hardship for local people and the animals with which they interact. If systems are in place that ensure efficient and effective response to drought-related problems, as was the case, for example, in Botswana in the 1970s and 1980s, the effects can be mitigated (Hay 1988).

The Sahel drought of the early 1970s brought the problems of the semi-arid savanna zones of Africa and the status of pastoral nomad populations to international attention. There have been different interpretations of the causes of what some described as an environmental disaster. On the one hand, there were those who argued that a major factor in the drought was the behaviour of local pastoral and agricultural peoples who kept too many animals on the range and used destructive cropping practices on land that was held as common property. On the other hand, there were those who suggested that it was the actions of outside aid agencies that played a significant role in the deteriorating environmental and social conditions in the Sahel. Still others maintain that the Sahel drought and related problems were a result of a combination of climatic, social, economic, and political factors.

Besides drought, Africa has also had to contend with livestock disease problems, including rinderpest (known as *seakhuma* in Setswana), which wiped out a sizeable proportion of eastern and southern African domestic livestock and wildlife in the 1890s. The rinderpest epidemic of 1896–7 led to a downward economic spiral that affected sizeable numbers of people in eastern and southern Africa (Van Onselen 1972). Such losses were especially problematic because a fairly significant percentage of the population depended on livestock and wildlife for subsistence and income.

The impacts of livestock diseases in Africa were reduced over time, thanks to the application of immunization and animal health protection efforts of governments and international agencies. In the 1940s and 1950s a vaccine was developed to deal with rinderpest. The development of a recombinant vaccine for rinderpest led to a wide-ranging vaccination effort in the 1960s which helped control the disease (Dobson 1995: 487–8). Until recently, there had been less success in dealing with other livestock diseases prevalent in Africa.

Over the past decade, the impacts of livestock disease have become important topics of discussion among policy-makers, governments, and scientists, in part because of the outbreak of mad cow disease (Bovine Spongiform Encephalopathy, or BSE) in England in November 1986. BSE is a fatal brain disease of cattle. It is believed to be caused by a self-replicating protein, a prion (Prusiner 1997). A major reason for the tremendous concern relating to BSE was that it is potentially linked to a human disease, Creutzfeldt–Jakob Disease (CJD). The efforts to control the spread of mad cow disease had significant effects on cattle producers in England as well as the British economy in general (Cherfas 1990). The export of British beef to other countries in the European Union (EU), for example, was banned until 1 August 1999. Of a national cattle herd of 10 million in Britain, 188,000 cattle at risk or affected by BSE were killed, and the milk from potentially infected cattle had to be destroyed.

The United States Department of Agriculture's Animal and Plant Health Inspection Service (APHIS) banned the import of live cattle and meat products from BSE-affected countries after 1989. But the United States had its own livestock disease-related problems to contend with. In the Midwestern United States, notably in Nebraska and Iowa, people became sick from eating ground beef infected by *Escherichia coli* in 1997. Some 30 million people per year are affected by food-borne diseases, including *E. coli* in the United States (Center for Agricultural Research and Technology 1994). There are complex interrelationships between livestock disease and human disease which have led to calls for improvements in meat inspection and measures to control the spread of livestock disease.

Africa has long had to contend with a variety of livestock diseases, some of which have outbreaks that coincide with droughts. This was seen, for example, in the early 1900s and again in 1933, when Foot-and-Mouth Disease broke out at the same time as a major drought occurred in Botswana (Botswana National Archives—BNA–S.312/18, S.312/19, S.312/20). Droughts are frequent in the arid and semi-arid parts of southern Africa and occur in two years out of five. According to Sandford (1977), severe droughts can be expected in one out of every four to ten years in southern Africa.

The serious droughts of the 1982–5 and 1992–3 periods in southern Africa brought into sharp relief some of the problems facing local people. In some cases, people were unable to plough, and they were forced to find alternative means of producing food and generating income. Fairly sizeable numbers of livestock died, thus reducing the chances for rural households to gain cash through sales of cattle or smallstock. Wild plant foods and game were also depleted seriously, thus reducing opportunities for people to use resources that often served as fall-back goods in stress periods. In order to counteract these trends, Botswana and other southern African countries mounted drought and disaster relief programmes.

The Southern African Development Community has identified drought as a major constraint for people residing in the region's countries. SADC has also identified three major livestock diseases that have had major effects on countries in the SADC region in the latter part of the 1990s: (1) Foot-and-Mouth (Hoof-and-Mouth) Disease, (2) cattle lung sickness (Contagious Bovine Pleuropneumonia, or CBPP), and (3) Newcastle disease, a disease that affects poultry. In 1995, the SADC region had 44 million cattle, 36 million sheep, 25 million goats, 36 million pigs, 1.7 million horses and donkeys, and 102 million poultry (data from the Southern African Research and Documentation Centre, SARDC).

In February 1995 it was reported that there was an outbreak of Contagious Bovine Pleuropneumonia in the /Kaudum (Xaudum) Valley in the north-western Kalahari Desert region of Botswana. The government of Botswana moved quickly to cope with the crisis, mounting efforts to eradicate infected livestock in the district and implementing a disaster relief and livestock-owner compensation program. Botswana Department of Veterinary Services and Animal Health (DVSAH) and other government and non-government personnel and soldiers from the Botswana Defence Force worked hard to contain the disease. Guards were posted hastily, and in at least one case a herd of cattle and its owner were stopped before the cattle could be taken across the Kuke veterinary cordon fence into the Ghanzi District.

As a means of preventing the spread of lung sickness, over 320,000 head of cattle were killed in Ngamiland in a relatively short period of time. In order to offset the effects of the cattle eradication, a program resembling the successful drought relief efforts of the Botswana government was implemented in northern Botswana, with some 44 million *pula* budgeted for the efforts (*Botswana Daily News*, 18 Nov. 1996, no 218: 1). Eventually, over 1.5 billion *pula* were expended on the livestock disease control and impact mitigation efforts. The reintroduction of livestock into the North West District began in April 1997, and was nearly completed at the time of writing this chapter. The losses of livestock due to eradication measures had significant socio-economic impacts on local people in Ngamiland and Botswana generally (Table 8.1).

This chapter addresses the efforts of the government and people of Botswana to cope with drought and outbreaks of livestock disease in the northern Kalahari Desert region. It outlines the ways in which drought and livestock disease-related relief activities have been implemented. The impacts of the 1977–8 Foot-and-Mouth Disease control efforts in Central and North West Districts are compared with those of the more recent Contagious Bovine Pleuropneumonia-related activities in Ngamiland. Particular attention is paid to the ways in which local populations in an environment some describe as marginal attempted to diversify their socio-economic systems in order to cope with changes brought about by drought, livestock disease, and the loss of access to livestock and livestock products.

Natural Resources and Livestock in Botswana

As is the case in many African countries, a sizeable proportion of the population of Botswana relies either directly or indirectly on natural resources to meet basic nutritional, economic, and social needs (Central Statistics Office 1976; Nteta, Hermans, and Jeskova 1997). Contemporary estimates indicate the contribution to the national economy of wildlife and other, non-domestic natural resources (e.g. fish, wild plant products) to stand at over P480,000,000 per annum (Ministry of Finance and Development Planning 1997). Soils and range resources are crucial to agriculture and livestock production, which together represent an important part of the economy of Botswana. While agriculture's share of the Gross Domestic Product (GDP) of Botswana declined from 42.7 per cent in 1966 to 4.1 per cent in 1994–5, agriculture

Table 8.1. Economic Contributions of Natural Resources in Botswana

Natural resources	Gross output (million *pula*)	Value added (million *pula*)	Employment (rounded)
Soils (for agriculture)	563.1	n.a.	84,000
Minerals (mining and quarrying)	4,859.0	n.a.	9,000
Rangeland (Grazing)			
Commercial	76.7	54.5	14,300
Freehold	17.7	0.4	1,200
Wildlife/Tourism			
Subsistence hunting	7.5–11.5	6.2	5,020
Photographic safaris	13.4	7.5	3,000
Hunting safaris	2.6	2.9	2,020
Game farming	0.1	n.a.	n.a.
Trade/processing	0.9	0.3	870
Hotels/tourism	1,839.9	n.a.	n.a.
Veld Products (Wild Plant)			
Subsistence	50.0	n.a.	n.a.
Commercial	4.4	2.2	100
Tourism-related	n.a.	n.a.	3,500
Fisheries			
Commercial	4.7	n.a.	900
Subsistence	2.6	n.a.	5,000–11,000
Forestry			
Commercial	4.7	n.a.	n.a.
Subsistence	2.6	n.a.	n.a.
Wood Fuels			
Urban	13.0	n.a.	400
Rural	16.6	n.a.	n.a.

Note: n.a. means not available. Figures are based on the most recent statistics available, 1995/6 and 1996/7.

Sources: The data in this table have been adapted from: Pearce, Barbier, and Markandya (1990: 158, Table 7.4); Harvey and Lewis (1990: 46, Table A3.1, and 76, Table 5.2); Pfotenhauer (1991); Nteta, Hermans, and Jeskova (1997); Ministry of Finance and Development Planning (1997); *National Accounts and Statistics Bulletins*, CSO.

continued to contribute significantly to the income and subsistence of many rural households (Harvey and Lewis 1990: 67–107).

The crucial variable affecting arable and livestock production in Botswana is rainfall. In general, the higher the rainfall the greater the biological production. There are other variables that also affect agriculture, including soil nutrients, pests, and disease. The Botswana Government and donor agencies have expended substantial sums in an effort to deal with these constraints. Sizeable sums have also been expended on dealing with the social and economic impacts of drought and livestock disease outbreaks (Sandford 1977; Hubbard 1986; Ministry of Finance and Development Planning 1997).

Livestock, and particularly cattle, play a variety of important roles in Botswana's economy and society, including serving as a kind of investment bank and providing a source of employment, food, and materials (e.g. hides for tanning). In 1980 exports of meat and meat products, live animals, and hides and skins brought in a total of P31,323,000. By 1995, these products earned a total of P217,913,000 (Ministry of Finance and Development Planning, 1997: 21). In 1994–5, the cattle industry accounted for 5 per cent of Botswana's exports, constituting 1.4 per cent of Botswana's Gross Domestic Product.

A major problem affecting livestock exports is the periodic outbreak of Foot-and-Mouth Disease, which has caused access to markets to be cut off, especially that of the European Union, an important one for Botswana beef (Hubbard 1986). Botswana has had the misfortune of having had cattle diseases affect the livestock industry as far back in time as records extend and presumably much longer (Falconer 1971a, b, 1972). Although Falconer (1971b: 74) maintains that there is no official record of animal disease in the Bechuanaland Protectorate prior to 1905, there are references to the 1896 rinderpest epidemic in the Botswana National Archives (see BNA file RC.3/2/1). The impacts of the rinderpest epidemic of 1896–7 were so severe that people in the major towns and villages of Botswana were forced to hunt and gather for their subsistence. Some of the people left the country, bound for the mines of South Africa. Others sold off their assets to raise cash with which to purchase food and other necessities.

Fear of the spread of rinderpest, which came from the north, led Bechuanaland Protectorate officials to erect one of the first veterinary control fences in Botswana, an east–west trending fence erected in the eastern part of the country, but by November 1896 rinderpest had spread to the southern borders of the country. In 1904 there was a threat of East Coast Fever near the borders of Bechuanaland, and a veterinarian was loaned to the Protectorate Administration in case it spread further. The first local veterinarian in Botswana was appointed in March 1905 (Jack Falconer, pers. comm. 1978).

One of the more serious outbreaks of FMD in the country was in 1933. Prior to that time, in 1931, there had been an outbreak in what was then Southern Rhodesia (now Zimbabwe) which was controlled by using an intranasal inoculation of virulent blood. This method, pioneered by a veterinarian named Bevan, was employed as a means of infecting susceptible cattle and in such a way building up their immunity to more virulent forms of the disease. A method not unlike that used by Bevan, known as apthization, was later used in Bechuanaland. This method consisted of using the material collected from the lesions on the tongue and feet, straining it, mixing it with a solution of glucose and penicillin and then injecting it into the animals. There was some question as to the efficacy of this procedure, as there was a chance that other animals in the area might become infected. This turned out to be the case when, in 1957, FMD broke out in southern

Botswana near an area where cattle had been heavily treated using apthization (Falconer 1972: 358). Because of the possibility of FMD spreading, it was decided in the early 1960s to use a system of annual prophylactic vaccinations; a serum for FMD that had been developed by the Animal Virus Research Institute at Pirbright, United Kingdom. Prior to 1948 it had not been known that the Southern African Type (SAT) strains of FMD were different from other strains, but once that was learned it was only a matter of time before a vaccine could be developed. The year 1965 saw the beginning of annual vaccinations of cattle against FMD.

An examination of data on FMD and other disease outbreaks indicates that they occur in Botswana every few years (see Table 8.2). The majority of the outbreaks have been in the northern part of the country, primarily in the Okavango, Chobe, Botletle, and Makgadikgadi areas. The incidence of FMD

Table 8.2. Livestock Disease Outbreaks in Botswana

Year(s)	Livestock Disease	Location	Reference(s)
1896–7	Rinderpest	Entire country	BNA RC.3/2/1
1933	Foot-and-Mouth	Entire country	BNA S.380/3, S.308/4, S.312/9
1934	Foot-and-Mouth	Botletle River, Central and Southern Districts	BNA 2/312/18, S.312/19, S.312/20
1937	Foot-and-Mouth	Palapye, Central Botswana	BNA S471/4; Falconer (1971*b*: 75)
1944	Foot-and-Mouth	Northern Botswana	BNA S.230/8/1–2
1947–9	Foot-and-Mouth	Northern Botswana	BNA S.230/9/1–4, S. 231/1/1–6
1950	Foot-and-Mouth	Chobe	Hedger (1968)
1957	Foot-and-Mouth	Southern Botswana	Falconer (1971*a*: 154, 1971*b*: 77, 1972: 358)
1958	Foot-and-Mouth	Ngamiland	Falconer (1971*a*: 154, 1971*b*: 77)
1960–61	Foot-and-Mouth	Northern Botswana	Falconer (1971*a*: 154, 1971*b*: 77)
1963	Foot-and-Mouth	Motopi, Botletle River area	Falconer (1971*a*: 154, 1971*b*: 77–8, 1972: 357)
1964	Foot-and-Mouth	Ngamiland	Falconer (1971*a*: 155, 1971*b*: 78, 1972: 357)
1965	Foot-and-Mouth	Central Makgadikgadi	Falconer (1971*a*: 155, 1972: 350)
1968	Foot-and-Mouth	Chobe	Falconer (1971*a*: 155, 1972: 354, 358)
1977–8	Foot-and-Mouth	Ngamiland and Central District	Falconer (1980), Hubbard (1986)
1995–6	CBPP (lung sickness)	Ngamiland	Ministry of Agriculture (1996)

Note: BNA stands for Botswana National Archives; the number following (e.g. S.308/3) is the reference file number used in the National Archives filing system.

in these areas must be, it was reasoned, associated with a common ecological factor. Looking at animals present in these areas, it was decided that
buffalo were likely to be the species responsible for carrying the disease.
Accordingly, the Department of Animal Health set about sampling a number
of buffalo in an attempt to test this hypothesis (Falconer and Child 1975;
Drager, Patterson, and Breton 1976). It was learned that buffalo carried three
types of virus, known as South Africa Types I, II, and III. Prior to this time,
no outbreaks of SAT II had been known to occur in domestic stock. It was
not until the outbreak in 1977 that SAT II was positively identified as being
responsible for the FMD occurrence in domestic animals.

There is some question in the minds of a number of people whether wild
species of game really are responsible for the outbreak of FMD in cattle in
Botswana, but there is no doubt that tests run on animals, and not only
buffalo, have shown there to be lesions and high levels of immunity, indicating infection with FMD (Condy, Herniman, and Hedger 1969; Condy and
Hedger 1974). After vaccination of the cattle herds in the Central District
affected zone it was realized that the vaccine was not as effective against SAT
II as it needed to be. The result was an inability to control FMD completely.
The cordon fence system and the patrols, however, managed to contain the
disease, so cattle from the northern part of the country could still be sold to
the Lobatse Abattoir. But the outbreak resulted in a decrease of livestock
sales throughout the country and a loss of millions of *pula* that Botswana
might otherwise have had. The threat of outbreak of FMD was recognized
by the Department of Animal Health, and a project was written to extend
the cordon fence system and to make the existing cordon fences smallstock-
proof. The National Development Plan VI. 1976–81 (Ministry of Finance
and Development Planning 1977) included a project (AH 12) to control FMD
through an extension of the fences, because the government had realized that
the vaccines had ceased to be as effective as formerly. One of the problems
of expanded fencing was that it not only prevented livestock movement but
also the movement of game animals, especially migratory ones. Falconer
(1972: 356) maintained that game changes its movement patterns away from
the fences so that they are not affected, a position that was challenged by
other researchers (Owens and Owens 1981, 1984; Hobbs 1981).

It is interesting to note that the Central Ngwato fence, reputedly patrolled
daily in 1978, was found to have long sections lying on the ground, and cattle
were moving freely back and forth across it even after the ban on livestock
movement was declared and the fences supposedly strengthened. When the
author reported this fact to government authorities, they claimed that this
breakage was due to 'a huge herd of eland, ranging in size from 400 to 1,000
animals'. Over two years of fieldwork in the region, no more than a maximum
of 70 eland in a herd were observed, and no animals were seen near the fence
in the period from October 1977, when FMD was first declared, to April
1978. Yet workers for the Veterinary Department went into the area and shot

dozens of eland. It is open to question whether elands or other game animals were actually responsible for the condition of the fences or whether they were in poor shape because of a lack of maintenance.

The impact of the 1977 FMD outbreak was all the more serious when viewed from a historical perspective. Botswana, plagued by periodic droughts (some of the most serious in the past few decades being those of 1933, 1947, and 1961–5, which, interestingly enough, correlate with FMD outbreaks), had a run of several years of abundant rains in the 1970s. This fact, coupled with favourable world beef prices and expanded veterinary and agricultural extension assistance, meant a tremendous increase in the size of the national herd to its largest number in history, something over 3 million head. At the same time there was a corresponding increase in the number of watering-points in the sandveld areas especially in the period following 1965 when 'Drought Relief Boreholes' were drilled. Many of these new boreholes were in the vicinity of the Makoba (Central Ngwato) quarantine fence (see Hitchcock, 1978: 161, fig. 6.3).

The 1970s saw a rapid increase in borehole drilling, in spite of the 1973 borehole freeze which was instituted prior to the launching of the Tribal Grazing Land Policy (TGLP) in 1975 (Republic of Botswana 1975) (for a discussion of this point, see Hitchcock 1978; Peters 1994). The additional watering-points resulted in an expansion of cattle further west in the country, including into the western parts of Central District, western and north-eastern Kweneng District, western Southern (Ngwaketse) District, north-western Kgalagadi District, southern Ghanzi District, and the South Ngami area of North West District.

There was a direct income loss to cattle-owners in Ngamiland and the northern part of Central District, as well as in the buffering areas of South Central District, and Ghanzi District. When the Lobatse Abattoir opened, only cattle from the Kweneng, Kgatleng, Kgalagadi, and Southern (Ngwaketse) Districts were taken. Central District was undoubtedly the hardest hit. In an analysis of the income effects of FMD, Mauco (1978) indicated that the only people who were directly affected by the disease were the wealthy or say 25 per cent, not the lowest 45 per cent who according to the Rural Income Distribution Survey (RIDS) own no cattle (Central Statistics Office 1976). He also noted that 'even the poorest households own lands'. Mauco went on to argue that 'the extent to which non-cattle owners will bear the loss (I contend) is minimal and depends on the Income Source Ranking of the lowest income group'.

Mauco's assessment was based on the results of the Rural Income Distribution Survey (Central Statistics Office 1976), which maintained that 45 per cent of the people in Botswana owned no cattle. However, a distinction must be drawn between 'owning' cattle and holding cattle. Many people have the use of cattle, for example as draft power or milk sources, while not actually owning them. The *mafisa* system is one in which individuals have the use of cattle over a period of time and essentially herd them themselves, getting the benefits of the meat, milk, and transport power. At the same time, many people borrow or hire cattle to use in their fields. While RIDS emphasized owning, the Agricultural Surveys of the Ministry of Agriculture (e.g. those conducted in 1970–1 and 1972 by the Agricultural Statistics Unit) stressed holding of cattle. A more realistic figure of those who did not have access to cattle was probably around 30–35 per cent.

Thus, the questions that need to be considered are: (1) what effects did the FMD outbreak have on those who did not own cattle? (2) what effects did the FMD outbreak have on those who did not hold cattle? and (3) what effects did the FMD outbreak have on non-stock-owners and non-stockholders (e.g. employees of cattleposts who have no livestock or *mafisa* holdings of their own, and dependents of those who work on cattleposts)?

The Rural Sociology Unit (now, the Socio-economic Monitoring and Evaluation Unit) in the Ministry of Agriculture was requested in early 1978 to carry out a survey of the effects of FMD in the Central and Ngamiland Districts. These surveys were carried out in January and February 1978, and two reports were produced (Ntseane 1978; Merafe 1978). In contrast to the view that FMD had minimal effects, these investigations indicated that the outbreak had drastic effects, particularly on the poorer sectors of the population. The ban on the movement of cattle, for one thing, meant that many of those people who had yet to plough their fields were unable to do so, since the cattle could not be taken from the cattleposts to the lands. Second, there was a general movement away from the villages to the cattleposts so that people could get access to milk which could not be transported from the cattleposts to the villages. This movement also resulted in a depopulation of the schools.

The economic effects of the outbreak included a reduction of incomes for people in the Ngamiland and Central District villages. People who made their money through the manufacture of beer or crafts had a hard time making sufficient sales. There was also a loss of employment related to the FMD outbreak, since store-owners and other employers cut back on their help. Even transport was affected; many people in rural areas move from one place to another by donkey, and with the ban on animal movement this strategy was no longer allowed.

Yet another problem identified in Ntseane's and Merafe's surveys related to purchasing and credit. People were forced to buy goods from stores since they could not get access to milk and meat. There were two problems with this strategy; first, the stores had little in the way of stocks, partly because of the tight money situation and partly because of the poor road conditions brought on by heavy rains. Second, since people had no money with which to purchase goods, they had to ask for credit, but generally the store-owners, hard-pressed themselves, refused to extend it to them.

In 1977–8 there were two opposing views in Botswana on the effects of

FMD. One view held that FMD was affecting only the wealthy, since it was they who owned the cattle and could not sell them. The other perspective was that the FMD outbreak was affecting everyone in the country, particularly the poor. Prices for goods increased, and thus people's purchasing power was reduced. Some of the people employed on cattleposts in the Western Sandveld region of Central District found themselves out of a job, or alternatively they were not paid for their work nor given food (Hitchcock 1978). One of the responses to this situation was that people resorted to alternative ways of making a living, including expanding their dependence on wild plant and animal foods, a strategy similar to that seen in drought periods in Botswana.

FMD, far from having a 'minimal' effect on the poorer sectors of Botswana's population in 1977–8, had a very marked effect. This was the case in the villages examined by Ntseane (1978) and Merafe (1978) and cattleposts in two of the commercial areas, the Western Sandveld of Central District and Lepasha, the Second Development Area (SDA) to the east of Sua Pan in Central District. In both of the commercial areas it was found that people suffered from the effects of the disease, particularly in terms of diminution of cash income, loss of employment, and lack of access to subsistence sources (such as milk, for example). The removal of cattle from cattleposts in the Western Central District and Lepasha to the quarantine camps at Makoba and Dukwe resulted in a loss of access to milk and draft power for cattlepost residents.

Employment opportunities for herders (*badisa*) declined in both the Western Sandveld and Lepasha areas, with the numbers of jobs being reduced on average by one-third to a half (Hitchcock, field data). While some people were able to obtain employment as fence guards, this did not make up for the loss of jobs for local people. The reason for this situation was that most of the fence guards who were interviewed in 1978 had come from outside the immediate area, mostly from the north or the west, from Mosu and Rakops, for example. Thus, there was competition for jobs in the areas affected by Foot-and-Mouth, something that exacerbated social tensions.

Because there was no off-take of livestock for sale, cattle-owners held back on the payment of their employees, and the people on the cattleposts said that they suffered greatly as a result. The one advantage of FMD to the local people in the Western Sandveld and Lepasha regions was that it meant that they could eat all the meat of a dead animal, something they were not able to do previously since they were obligated to take the meat of the animal to the owners, most of whom lived in villages outside the commercial zone. A number of cattlepost employees mentioned that this increased access to the meat of dead livestock was the one benefit that came out of the FMD situation.

The people on the cattleposts suffered other consequences as a result of the government rules about actions that were not allowed in the affected areas. There was punishment meted out by fence guards to individuals who took their animals too close to the cordon fences or who carried meat or milk across fences. There was also stepped-up monitoring of rural areas by Department of Wildlife and National Parks and Department of Animal Health personnel, and one result was an increase in the number of arrests of people for contravening the ban of hunting and moving of animal products (both domestic and wild).

It is ironic that while there was a ban on hunting by local people, this was not the case for the veterinary teams, who had the right to shoot any animals that they considered a threat to the veterinary cordon fences or to the cattle in the affected areas. During fieldwork in 1978, scenes were witnessed where at least a dozen eland and hartebeest had been killed by veterinary teams and the only parts taken from them were the livers. When it was pointed out to the Department of Animal Health workers that the taking of livers across cordon fences was illegal, they said that they had government permission to do so, something that remains unconfirmed.

The hunting ban not only had an effect on local nutrition but also on the relationships between local people and government officials, which deteriorated rapidly, especially in the areas close to the veterinary cordon fences where people were arrested for crossing the fence after allegedly procuring game meat. These tensions were especially obvious in confrontations between groups of local people and game scouts that sometimes resulted in physical altercations.

Yet another effect of the FMD outbreak was that the borehole pumps were turned off by owners as a means of saving money on diesel. The borehole engine at Uwe-Abo (Central District, see Ch. 5), for example, was turned off when the cattle were taken to the Makoba Quarantine Camp in January, 1978. This situation caused local people to go thirsty. Some of the people in the Western Sandveld resorted to exploiting roots or melons (e.g. *Citrullus lanatus*) (Hitchcock 1978). The problem was that the presence of large numbers of cattle had resulted in a serious reduction in the availability of melons and plants with large moisture-bearing roots. There was also increased competition between people from villages and lands areas with those residents of cattlepost areas for wild plant resources. The result was increased hardship for those groups who relied on wild resources for moisture, food, medicines, and income.

There are a number of conclusions that can be drawn from the assessment of the 1977–8 FMD outbreak impacts. The declaration of FMD had substantial effects on the poorer sectors of Botswana's population not only in the towns and villages but also in the cattlepost areas. These effects included: (1) the loss of employment; (2) the loss or reduction of payments for labour; (3) the reduction in subsistence resource availability as a result of the cattle being moved to the quarantine camp and the ban on moving milk and other livestock products; (4) the loss of draft power and the consequent inability for people to plough, resulting in a reduction in field sizes and agricultural yields; (5) the loss of the opportunity to sell cattle for income-generating purposes; (6) the loss of access to water in some cases; (7) the failure of credit being extended to people by trading store owners; (8) an increase in prices for goods; (9) a reduction in school attendance; and (10) the increased rates of arrest of people for breaking rules concerning moving livestock products, herding cattle too close to a veterinary cordon fence, and hunting. Overall, the quality of life for people on cattleposts, in lands areas and in villages, declined as a result of the FMD outbreak of 1977–8.

The Botswana government initiated a relief operation that included what was known as the P50 Scheme, in which individuals were able to get credit through the Co-operatives and the Botswana Livestock Development Corporation (BLDC). The government also abolished payment of primary school fees in all the affected areas, while payment of secondary school fees was deferred. The loan scheme, while considered by some to be too little too late, was instrumental in providing cash for purchases of food and other goods at the local level. Some people were able to pay off part of their National Development Bank loans with the money that they received, while others were able to take care of some or all of their outstanding debts with local traders. The problem with the scheme was that it was set up initially to provide credit to people who had cattle already; thus, those who did not own cattle were not able to benefit directly from it. There was also a stipulation that the loan would be made only to those who could pledge oxen as collateral; those who had heifers felt that they were left out (Merafe 1978).

The P50 Scheme was viewed by local people as being inequitable because it provided support to cattle-owners and not to either arable farmers or non-stock-owners. Overall, the scheme was useful in the eyes of those who received funds but even they had their reservations about the amounts of money made available. The other issue that local people had with the scheme was that the implementation by the BLDC and the co-operatives was not as effective as it might have been. They felt that other institutions in government, including the District Councils and the Remote Area Development Programme, should also be involved (Yvonne Merafe, pers. comm. 1978). There should be greater co-ordination and co-operation among government- and district-level organizations, and communities should be allowed to have a greater say in the planning and decision-making process.

Dealing with the CBPP Outbreak in Ngamiland

The relief operations that Botswana has mounted in the past to deal with drought and livestock disease outbreaks have generally been viewed as quite effective. From the standpoint of those who had no cattle or those who were working on cattleposts, however, the programs were seen as largely aimed at the better-off sectors of the population. The erection of fences was seen by some people as being a useful way to prevent the spread of livestock disease, but they were also perceived as being harmful to wildlife (Hobbs 1981; Owens and Owens 1981, 1984; Albertson 1998).

After the presence of CBPP was confirmed in 1995, wide-ranging vaccination campaigns were employed using TI-SR CBPP vaccine produced by the Botswana Vaccine Institute (BVI). Guards were posted along fences in order to prevent movement of livestock and livestock products. Plans to allow the repatriation of Herero livestock from Ngamiland to Namibia were shelved. A major difference between the 1977–8 FMD outbreak and the CBPP outbreak of 1995 was that in the latter case, the Botswana government opted to kill all the cattle, regardless of whether or not they were infected with the disease. Also, a much more concerted effort was made to provide timely cash and in-kind compensation to those who lost cattle.

Some of the impacts of the CBPP campaign and relief efforts in Ngamiland can be seen in the case of /Xai/Xai (Cgae Cgae), a community of some 388 people in the western remote zone of the district, next to the Botswana–Namibia border (see Figs. 8.1 and 8.2). At the time the decision was taken by the government of Botswana to dispatch all the cattle in the district to halt the spread of CBPP, there were several hundred cattle at /Xai/Xai. Cattle-ownership was skewed, with only a third (36%) of the households having cattle of their own. The cattle were grazed in areas that have now been designated as Community-Controlled Hunting Areas (CCHAs) NG 4 and NG 5, which together cover an area of 16,966 km^2. Some of the people from /Xai/Xai worked at a nearby cattlepost, Xhaba (Botshelong), where there were over 1,000 head of cattle at one stage in the early to mid-1990s. Overall, in 1991, there were some 21,000 cattle in western Ngamiland, a figure based on an assessment of numbers of animals vaccinated for FMD (Smit and Kappe 1992: 30).

By mid-1996, there were no cattle left in western Ngamiland, the herds having been eradicated by teams made up of government workers including people drawn from the Botswana Defence Force, the Ministry of Agriculture, and the North West District Council. After the cattle were killed, people at /Xai/Xai diversified their strategies. Some of them engaged more intensively in foraging for wild foods, especially wild plant foods such as *morama* (*Tylosema esculenta*) and *mongongo* (*mokongwa, Ricinodendron rautanenii*). There were also people who opted to engage more extensively in craft production and sale, making bead necklaces and leather carrying bags.

Other economic activities that people pursued at /Xai/Xai included hunting both for subsistence and for commercial purposes, with some people selling hunting products to local buyers, to tour groups that visited the area, and to the Botswana Defence Force soldiers who were patrolling the border. A number of the residents of /Xai/Xai joined a craft organization, !Kokoro Crafts, which in 1995–6 earned P13,500. Approximately twenty people took part in eco-tourism activities with !Kokoro Safaris, earning some P8,000 (Ruigrok 1997).

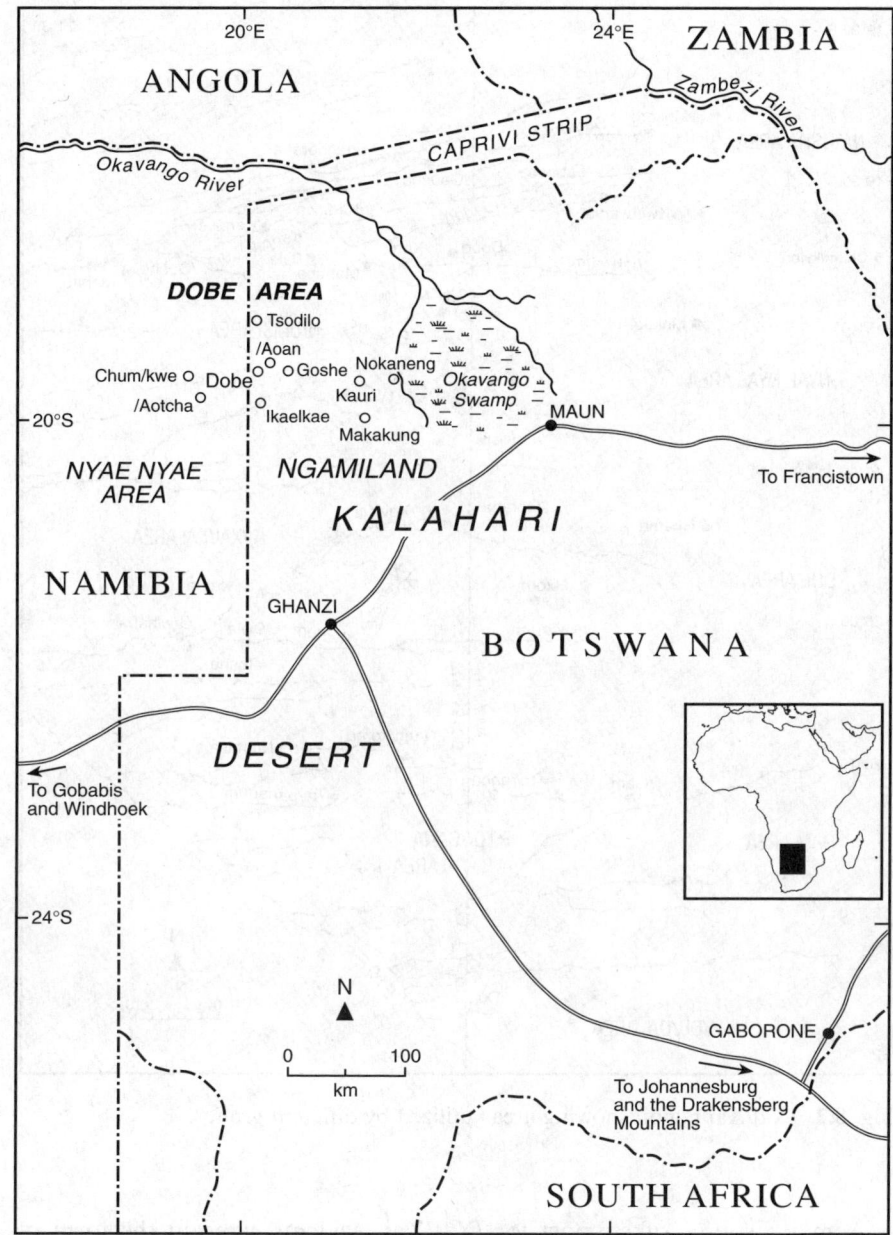

Fig. 8.1. General location of the area referred to in the text and principal names.

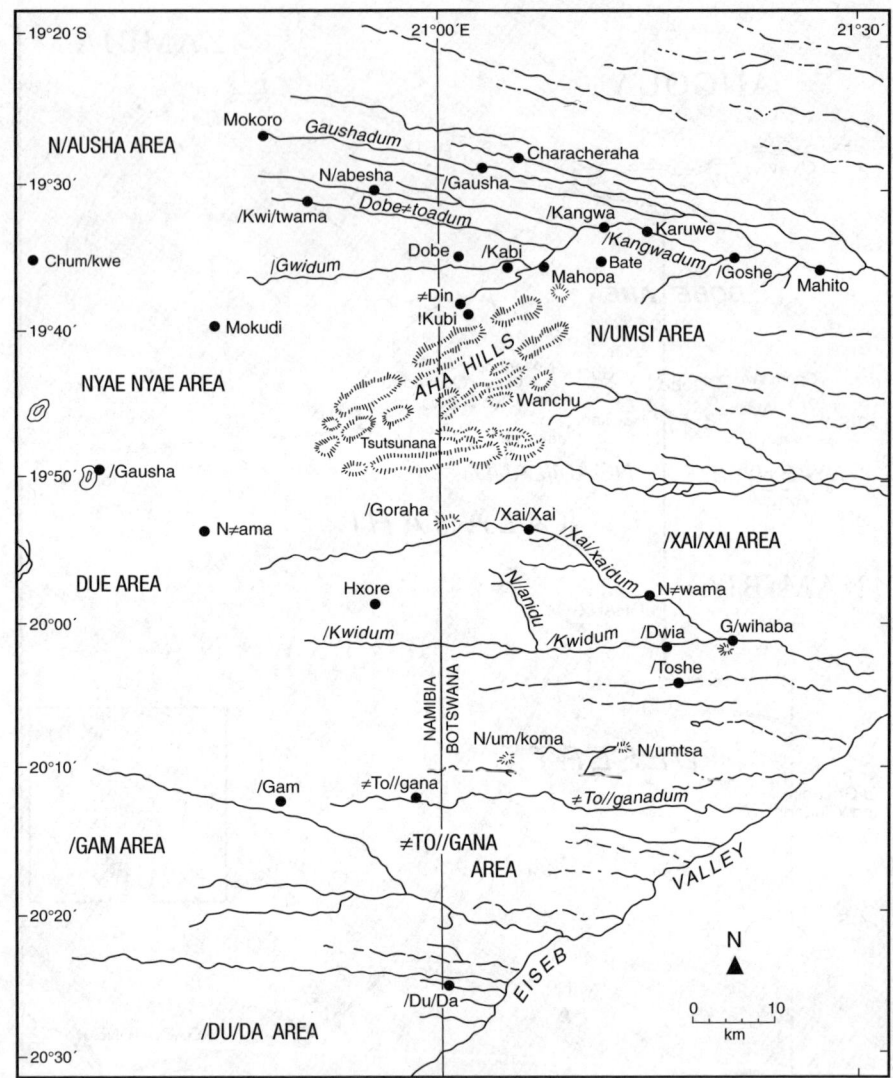

Fig. 8.2. /Xai/Xai region showing areas utilized by different groups.

A major source of support for /Xai/Xai residents came in the form of employment by government, which included road clearing for the North West District Council and the Roads Department. Six people worked for the Geological Survey, while two worked for the Natural Resource Management Advisor (NRMA) from SNV, the Netherlands Development Organization. Two other people worked as borehole pumpers, while one worked for the

Tribal Administration (the headman), two worked for the health department of the Council, and one served as a warden for the Tawana Land Board. A total of eighty people were employed, the majority of them being paid to do road clearing.

Botswana's effective drought relief programme was expanded upon in the CBPP relief efforts. Not only were local people able to get work through labour-based relief and development projects, but they also received rations similar to those provided to destitutes by the District Council, including maize, sorghum, beans, and oil. The amounts provided to people depended on household size. As one development worker put it, there was so much food in some of the households that they had to sleep outside. Another person at /Xai/Xai said that she was happy for the food and the jobs, but that she would prefer to have her cattle. Yet another woman suggested that the CBPP relief food would eventually run out and that people would be 'back where we started from'. The majority of the people were definitely pleased that the government had provided them with assistance, and they looked forward to receiving their new cattle once the restocking exercise was implemented at /Xai/Xai.

Overall, the lessons from the 1977–8 FMD relief operation were not lost on the Botswana Government and the North West District Council. Efforts were made to provide significant amounts of cash and relief food relatively quickly in order to offset the negative effects of the CBPP cattle eradication. A variety of government institutions and non-government organizations took part in the relief and development efforts in Ngamiland. Members of the /Xai/Xai community were given the opportunity to assess the impacts of the newly constructed Setata Fence, a veterinary cordon fence constructed in the western Ngamiland region from the Namibian border to the Buffalo Fence to control the spread of CBPP. It is interesting to note that, as in 1977–8, the opinions of local people had no impact on the government's decisions on the construction, location, and type of cordon fence in spite of its negative impact on wildlife in the western Ngamiland region.

The decision of the government to erect new cordon fences in the area was greeted with dismay by some of the people in western Ngamiland. While there was some variation in the views that people had, there was a general sense that the establishment of the fences was the first step towards the enclosure of the open rangelands of the north-western Kalahari. Some people noted that the fences were contributing to the losses of wildlife, while others said that the fences were going to limit the movements of their domestic animals with potentially harmful effects in drought periods.

In the mid-1990s, the Botswana Defence Force considered establishing a military base in the western Ngamiland area. Although the base has yet to become a reality, the numbers of military personnel in the area has been fairly high. One impact of the increased presence of the military was larger numbers of arrests of local people for engaging in subsistence hunting. In one case, in June 1995, the Anti-Poaching Unit arrested seven men for hunting illegally and confiscated all their equipment and the horses and donkeys they were using (Hitchcock and Masilo 1995; Hitchcock *et al.* 1996). The Botswana Defence Force also raised questions about tourists being brought to /Xai/Xai by professional tourism groups in 1996. The response to these concerns on the part of some local people was that the military was attempting to impose restrictions on their livelihoods.

The loss of livestock had significant socio-economic impacts on local people in western Ngamiland and the northern Kalahari in general. Some people fell back on foraging, going into the bush to collect wild foods. Others crossed the border into Namibia to live with relatives and friends. Still others moved to the towns around the Okavango Delta such as Gomare, Nokaneng, and Maun where they sought employment or lived with relatives. Access to milk was reduced, which affected the nutritional well-being of the local populations in western Ngamiland. The lack of livestock also had social impacts, with local people not able to pay bride-price (*bogadi*) in the form of cattle. There were also difficulties because people had to rely on donkeys for pulling ploughs, something that affected agricultural production.

In order to counteract the negative impacts of the cattle eradication, the Botswana government established a cash compensation programme for those people whose livestock were killed. The government paid some 500 *pula* per animal for 70 per cent of the animals belonging to a cattle owner, with the other 30 per cent of the compensation given in kind. People were also given food, including maize, sorghum, beans, and oil. This food was distributed fairly widely, with some of it being circulated to relatives across the border in Namibia or sold to people who were better off, including government officials working in the region.

Botswana has mounted a number of successful drought relief feeding activities over the past several decades (Hinchey 1979; Holm and Morgan 1985; Hay 1988). Three out of five people in Botswana were receiving food aid at the peak of the feeding programme in the mid-1980s. In the latter part of the 1990s, food was provided to the entire population of Ngamiland, with the exception of some of those people living in urban areas.

In order to cope with the CBPP crisis, the Botswana government initiated a multipronged effort that included commodity aid, cash-for-work programmes, compensation for lost livestock both in kind and in cash, and suspension of loans at the National Development Bank. The government expanded its supplementary feeding programmes for school children from 191 days a year to 365 days. Assistance was also provided for medically selected (i.e. specially targeted) malnourished children. The agencies responsible for these activities included the Food Resources Department (FRD) of the Ministry of Local Government, Lands, and Housing and the North West District Council. The programme was almost entirely dependent on food aid from donors in the past, but under the CBPP programme, the government

used Domestic Development Funds (DDF). Overall, according to government analysts, the programme was implemented and executed smoothly,
though there were some initial logistical difficulties. There was also some
leakage (i.e. some of those who were supposed to get specified amounts of
food did not get it, or got only a portion of what they were supposed to
receive).

One group that received special assistance was that defined as Remote Area
Dwellers (RADs). Remote Area Dwellers were provided with a full ration
that was intended to cover all their food requirements. No attempts were
made to differentiate among the people in need within the RAD target group.
All of the people in this category were regarded as being seriously affected
by the livestock disease control efforts, and they were seen as having few alternative sources of income. It should be noted that in spite of the availability
of food, possible micro-nutrient deficiencies may have resulted from the composition of the ration. This is the case, for example, with a lack of sufficient
Vitamins A and C. The usual sources of these micro-nutrients are products
such as wild plant foods. These goods were much in demand as a result of
increased numbers of people engaged in foraging.

The rural economy in Botswana is both varied and complex (Central
Statistics Office 1976; Chernichovsky, Lucas, and Mueller 1985). Rural people
tend to spread their risk; in other words, they diversify their subsistence and
income sources (Hitchcock 1978). According to the RIDS survey of 1975–6,
the relative importance of income in kind is greater among poorer households than among those that are better off. There is thus an inverse relationship between income in kind and income level. Among lower-level income
groups, there is a much greater degree of dependence on transfers, both
private, such as mine labour remittances and public, such as the government
relief provision. Among the poor, including RADs, there is a relatively high
level of dependency on government welfare and assistance programmes. This
must be viewed in a balanced way, however. Currently there are few economic
opportunities available to people in rural areas other than tourism and craft
production.

In spite of the difficulties facing people due to the loss of livestock, there
were some positive spin-off effects. One of these was that the grazing and
browsing resources had a chance to recover, as did the wildlife populations.
The complications of the cattle eradication campaign also had the effect of
bringing people together who in the past were at loggerheads over issues such
as land use. The Herero in /Xai/Xai, for example, who kept fairly sizeable
herds of livestock, wanted access to the G/wihaba and Aha Hills areas, which
traditionally belonged to Ju/'hoansi *n!ore kxausi*, the 'owners' of the land
around the two sets of hills. The Ju/'hoansi San and the Herero at /Xai/Xai
began to talk about forming a quota management committee (QMC), which,
once established, could be granted the wildlife quota for the CHAs NG 4 and
NG 5 in western Ngamiland under new Botswana Government legislation

relating to community-based natural resource management (Republic of Botswana 1986, 1998).

In October, 1997, the people of /Xai/Xai formed the /Xai/Xai Tlhabololo Trust, and as a result they were able to obtain the wildlife quota from the Department of Wildlife and National Parks. Some of this quota they set aside for community use, allowing local hunters to take a portion of the animals available. The rest they pledged to commercial use, and they leased out the rights over that quota to a private safari operator. According to Motshubi (1999: 46) the people of /Xai/Xai were able to generate P40,000 from hunting, P20,000 from photo-tourism, and P20,000 from craft production in the first year of operation. Some of these funds were derived from individuals who camped in NG 4 (see Fig. 8.3), who were required to pay a camping fee of P10 per night to the trust.

A major concern of local people at /Xai/Xai related to central government plans for the expansion of the national monument around the caves at G/wihaba (Drotsky's Caves). The G/wihaba Caves had been declared a National Monument under the Monuments, Relics, and Antiquities Act (No. 15, 1970; see *The Laws of Botswana*, Ch. 59.03). In 1995, the National Museum, Monuments, and Art Gallery told the people of /Xai/Xai that they wanted to establish a conservation zone around the G/wihaba and Koanaka Hills because they contained resources of scientific importance. Discussions with people from /Xai/Xai revealed that there were a variety of sentiments about such plans. Some people wanted to see a conservation area established because they thought that it would increase the number of tourists coming to the area. Others were opposed to the idea because they felt it would infringe on their hunting and gathering rights. Still others wanted to have a conservation zone around the hills on the conditions that they were allowed to continue to use the resources there, and that they received a portion of the gate receipts (Hitchcock *et al.* 1996). Several people said that they would be willing to support the establishment of a conservation zone by the National Museum only on the condition that the Botswana government passed a new piece of monuments legislation that allowed for local people to be granted a portion of the benefits that were produced. It was clear, therefore, that the establishment of new legislation on conservation is viewed as an important aspect of future efforts to ensure the well-being of local people and habitats in the Ngamiland region.

There were efforts in the late 1990s to map the traditional territories (*n!oresi*) areas surrounding communities in western Ngamiland so that people could establish claims to them. Such efforts became especially important in the light of an announcement in mid-1999 that there was a likelihood of the government of Botswana establishing commercial cattle ranches along the Botswana–Namibia border. Thus, the efforts to take advantage of Botswana government plans to promote the formation of community trusts under its community-based natural resource management programme assumed even

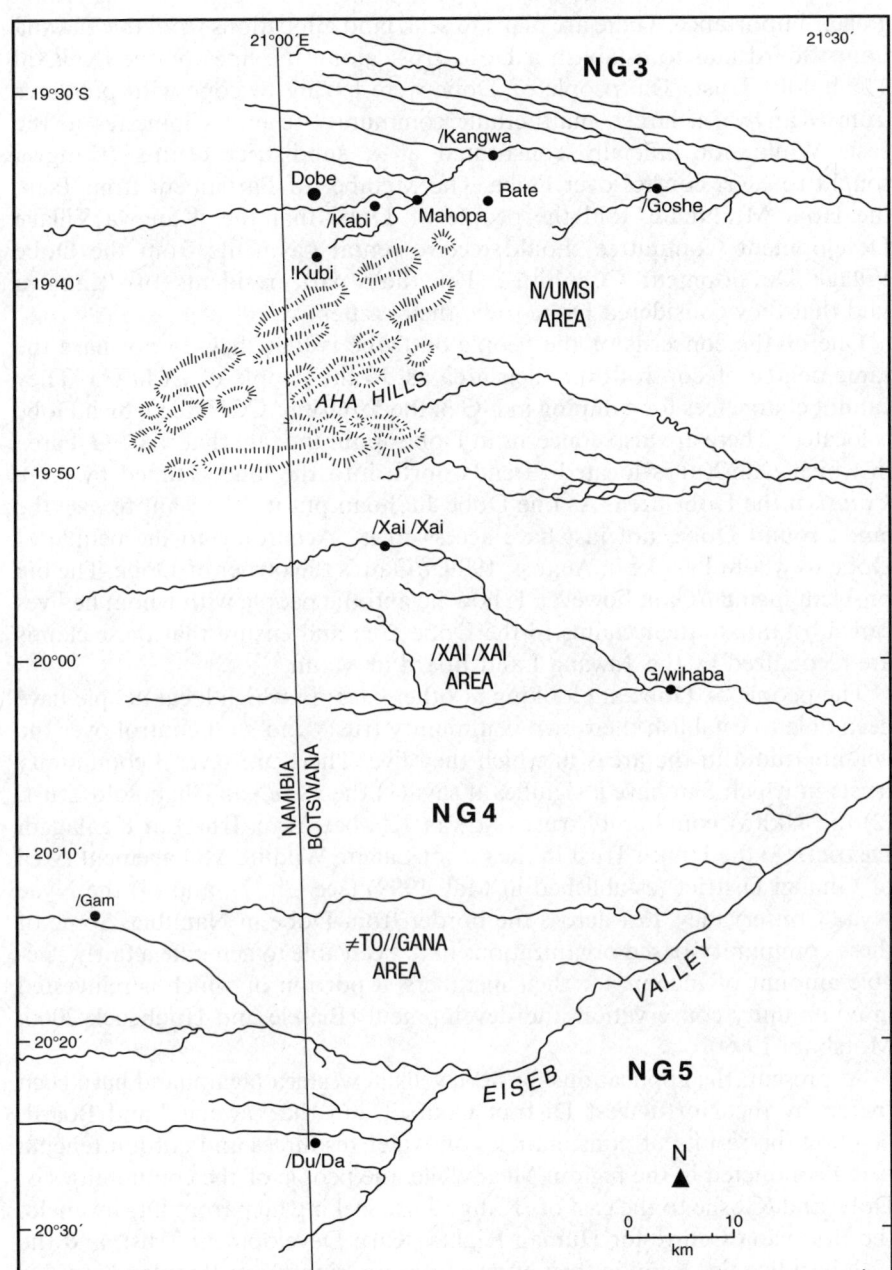

Fig. 8.3. Controlled hunting areas in the /Xai/Xai region.

greater importance. There are plans to seek land allocations from the Tawana
Land Board and to establish a Dobe trust along the lines of the /Xai/Xai
Tlhabololo Trust. The people of Dobe were having to cope with pressures
from /Kangwa, a larger, multi-ethnic community several kilometres to the
east. While not officially constituted as a subdistrict centre, /Kangwa
sought to exert control over Dobe. The Member of Parliament from Tsau,
the Hon. Mr Nkate, told the people of Dobe that the /Kangwa Village
Development Committee should receive rental payments from the Dobe
Village Development Committee. For their part, residents of /Kangwa
said that they considered Dobe to be their cattlepost.

One of the concerns of the people of Dobe is that they do not have the
same degree of control over their area as do the people of /Xai/Xai. They
cannot charge fees for camping in NG 3, the 5,760 km^2 CCHA in which Dobe
is located. There are also concerns in Dobe about the fact that the NG 4 area
in which /Xai/Xai is located extends north into the land claimed by *n!ore
kxausi* in the Dobe area. As one Dobe Ju/'hoan put it, 'We want to *own* the
land around Dobe, not just have access to it.' According to the people of
Dobe to whom I spoke in August, 1999, //Gau is the owner of Dobe. The big
problem facing //Gau, however, is how he and the people with whom he lives
can substantiate their claims to the Dobe area and ensure that these claims
are recognized by the Tawana Land Board in Maun.

The people of Dobe are looking at other cases in which local people have
been able to establish their own community trusts and gain control over the
wildlife quota in the areas in which they live. There are several community
trusts in which San have a significant say: (1) the /Xai/Xai Tlhabololo Trust,
(2) the Ukhwi community trust (Nqwaa Khobee Yeya Trust) in Kgalagadi
District, (3) the Huiku Trust in the Groot Laagte Wildlife Management Area
of Ghanzi District (established in May 1999) (see Ch. 7), and (4) the Nyae
Nyae Conservancy, just across the border from Dobe in Namibia. Some of
these community-based organizations have been able to generate a fairly sizeable amount of income for their members, a portion of which is reinvested
in community conservation and development (Biesele and Hitchcock 2000;
Motshubi 1999).

At present, the applications for new wells in western Ngamiland have been
frozen by the North-West District Council and the Tawana Land Board,
pending the results of consultancies on water resources and cordon fencing
being conducted in the region. Meanwhile, the people of the communities of
Dobe and /Goshe to the east of /Kangwa are seeking help from Ditshwanelo,
the Botswana Centre for Human Rights, Kuru Development Trust, and the
Kalahari Peoples Fund in their efforts to secure rights over their land. At this
stage, it is unclear as to what will transpire with the land claim efforts in
western Ngamiland, but there is no question that people are working together
in an effort to obtain clear title over their lands and resources, raise their
incomes, enhance their well-being, and improve the quality of their lives.

Strategies for Coping with Uncertainty

People in the northern Kalahari, like those in other parts of southern Africa, are flexible, resilient, and innovative in their approaches to solving environmental problems and overcoming constraints. When a drought or some other kind of stress situation commences, human adaptive responses are set in motion. An almost universal response of Kalahari peoples is to diversify their strategies, exploiting a variety of different kinds of resources and taking advantage of economic opportunities. Another common response is to move, either away from places afflicted by drought or disaster, or to places where resources are available, such as towns. The formation of social alliances and the use of buffering strategies such as food storage have also proved to be very useful in dealing with disasters and uncertainty.

Natural resource management among Kalahari populations is based on extensive and environmental knowledge and experience. They monitor the environment carefully, looking for patterns and trends. Data are obtained on the state of the ecosystem, and information on the distribution, abundance, and types of resources are shared actively among community members. The generalized subsistence system of groups (e.g. those who are foragers or agro-pastoralists) meant that they do not place too much pressure on a few plants and animals. As the amounts of specific desirable resources declined, people will shift to alternative resources. Not only do people map themselves onto resources through movement, but resources are mapped onto people through sharing and exchange systems. People are also mapped onto people through kinship and social alliances, including marriage and trade partnerships, and exchange relations.

The landscape on which local people reside and earn their livelihoods is not simply undifferentiated space; rather, the land is divided into tracts that comprise the basic subsistence and residential areas of local groups. They have long-standing rights to these areas, and in many cases they have ideological connections to them as well. Resource areas, which some social scientists have called 'territories', contain a number of different kinds of resources, including water-points, wild food plants and medicines, wild animals, trees and shrubs for shelter, fuel, and building materials, and materials such as termite earth and ochre used for construction and for adornment (Hitchcock 1978; Silberbauer 1981). Territories are named areas that have boundaries that people generally are familiar with. In order to enter another group's territory, permission must be sought from the traditional land-holders, usually those individuals who have resided there the longest. The rights to these territories are usually inherited from one's parents, although there were cases in which people obtained customary rights to land through moving into an unclaimed area and establishing occupancy rights or seeking rights from traditional authorities.

One of the strategies for coping with drought and socio-economic uncertainty employed by local people was to request permission to move to another group's territory which had sufficient resources to sustain a larger number of people. Usually people asked permission to visit the territories of people with whom they already had social ties, such as those created through marriage (affinal ties) or ones that came about through trade partnerships. In most instances, if the territory 'owners' felt that there were enough resources available in their area, they gave permission for the other people to enter. There were cases, however, when permission was refused, especially in times of extreme drought. This was said to have been the case in the central Kalahari region of Botswana in 1933, for example, when a lengthy drought saw large areas affected, so much so, according to informants, that even the large trees along dried-out rivers were destroyed.

In times of drought and disaster the people of the northern Kalahari some-times sought to prevent the overexploitation of specific species through the use of taboos. These taboos included restrictions about the kinds of species that one could collect or consume. There were limits placed on the use of certain species; some nut- and fruit-bearing tree species such as *marula* (*Sclerocarya caffra*), for example, were not supposed to be cut for firewood. Jural rules about resource use also varied according to age, gender, and, in some cases, individual personal characteristics. Some animal parts were reserved for elderly people. Certain species of medicinal plants were supposed to be used only by people defined as traditional healers. There are also specific items that were to be used exclusively by people who had the ability to go into trance and conduct healing ceremonies.

Another way that northern Kalahari populations buffered themselves against possible resource scarcity in drought periods was to store food. When game was killed, some of the meat was dried by cutting it into strips and hanging it in trees, turning the meat into what in southern Africa is known as biltong. There were also cases in which sizeable amounts of wild plant resources were obtained and kept in people's houses or hung in bags in trees. In some cases, melon rinds were cut into strips and dried for later consumption.

In order to cope with extreme drought conditions, people would scavenge the meat of animal carcasses. One problem with this strategy was that the condition of the carcass was often such that the nutrient value was low. Another was that in some cases diseases such as botulism were caught from the carcasses of dead animals. Sometimes people collected bones and sold them to farmers who used them to feed their cattle or sold them to bonemeal factories to raise income. It should be noted here that the sale and use of bonemeal has been restricted in the United Kingdom as part of the government's efforts to control the spread of mad cow disease. The list of Specified Bovine Offals (SBOs) that were banned for use in the livestock and food industries was expanded as a result of the BSE crisis.

Local people employed creative social strategies to deal with potential resource problems. One such strategy was to establish a fictive kinship bond through exchange systems (Wiessner 1977). The social ties that were created by these exchanges were important because they facilitated the sharing of information about the state of the environment in various areas. They also helped establish a widespread network of fictive kin who had mutual rights and obligations. These people could then be contacted if a group or individual was under duress in the hopes that they might be able to accommodate them by providing food, water, and a place to stay.

There were cases in extreme drought periods when people experimented with new kinds of plants, essentially diversifying their diets. It was at these points that people resorted to eating *Acacia* gum, tree bark, and roots with which they were not familiar. This could be a risky strategy since some of these plants had high levels of secondary compounds that were toxic, and the number of plant poisoning cases often rose during drought periods. In order to get around this problem, various methods were employed to process these plants, including soaking them for long periods in water.

A common strategy of Batswana is to move their herds or flocks to places where water and grazing are available. In some parts of the northern Kalahari, people would send their animals to other areas in the company of herdboys (Schapera 1943). Families would sometimes split up into production units, with women and young children staying at the homestead while the men went to the mines, towns, or remote grazing localities.

A social response of people to drought in the northern Kalahari and Africa generally is the migration of people to other places in order to seek employment or to find relatives and friends upon whom they can depend for support. This strategy can work for a while, but the longer the drought lasts the more difficult it becomes for the hosts. A problem with the migration option is that the numbers of jobs and resources tend to decline in drought periods, thus increasing the competition among individuals and groups.

An ingenious coping strategy of people in southern Africa is to engage in the long-term loan of domestic animals to other people, either kin or non-kin, who then manage those animals for them in exchange for the use of livestock products such as milk and dung. This livestock loan system, which is known as *mafisa* in Botswana and *sisa* in Swaziland, is sometimes structured in such away that the people to whom the animals are loaned are given a young animal (e.g. a calf) at the end of the year in exchange for their labour. This social custom enabled livestock-owners to distribute their animals among a number of different people in different places, thus reducing their risk.

Northern Kalahari farmers, especially women, engaged in a number of different kinds of resource management strategies, drawing on their extensive indigenous knowledge, in order to cope with uncertainty. They maintained seeds of a variety of different kinds of crops, selecting these carefully for specific characteristics such as drought- and pest-resistance, types of soils in

Robert Hitchcock

which they grow, and taste. They staggered their planting so as to ensure that the crops got moisture during the highly variable rainy season. They also planted their crops in a number of different places, fragmenting their fields intentionally so as to maximize the chances that they would get at least some moisture so that they could obtain a minimum crop yield.

Farmers experimented with innovative farming and soil and water conservation techniques. They erected water-harvesting systems, built retaining walls, and ploughed fields in the winter so as to increase the chances that moisture would percolate into the soil when the rains came in the summer. Farmers used a number of innovative methods for restoring soil fertility, including adding green mulch to the soil and composting. They attempted to monitor their fields more closely in growing periods, in part to prevent animals such as small antelopes and baboons (*Papio ursinus*) from getting into them. They sometimes place small traps (e.g. deadfalls, snares) at strategic points in the fences around the fields in the hope that they could catch the animals that were raiding their crops at night.

Farmers attempted to control the degree to which crucial resources, such as stubble in the fields, were exploited. They limited the number of cattle that could come into the fields, or they restricted the timing of their use of the stubble, which they realized was a highly nutritious feed source. A coping strategy of many people in southern Africa is labour intensification, where people expand their work effort. One way that they did this was to spend more time in the fields. They also brought in additional labourers such as children to assist in agricultural, domestic, and off-farm activities. It was not unusual in drought periods to see very young children watching over the fields or herding animals and collecting water, firewood, and crop residues to feed livestock.

In order to buffer themselves against nutritional deficiencies in drought and other stress periods, northern Kalahari populations collected wild plants, some of which they kept for home consumption and others of which they sold. Some people also collected wild insects and occasionally caught small wild animals, some of which they consumed and parts of which they sold (e.g. to traditional healers who would use the animal parts in medicine).

It is interesting to note that there may be a differential response of females as opposed to males to drought problems. In many parts of Africa it is women who are involved extensively in agricultural work, whereas men tend to do more of the herding of livestock. As a result, when a drought occurs, women tend to lose agricultural employment opportunities more readily than men. Women also tend to have fewer options in terms of going to town to seek employment because of their domestic responsibilities. In order to cope with this situation, women engage in such activities as making and selling crafts and doing domestic work for other people.

Southern African populations have coped with drought problems by depending on their governments or aid agencies for support while pursuing their traditional socio-economic activities as much as they can given prevailing environmental and social conditions. Drought relief programmes that were implemented in southern Africa saved the lives of many people. At the same time, they contributed to what some analysts have described as a 'dependency syndrome', in which people gave up their regular activities and became reliant on handouts. In order to avoid this problem, governments and international relief agencies established labour-based rural development projects in which local people are given cash for work such as removing brush and stumps from agricultural fields or clearing trees and shrubs in order to make roads.

The drought relief programmes mounted by the Botswana government and international agencies have been judged as relatively successful in averting the disastrous consequences that affected other African countries (Holm and Morgan 1985). Efforts were made not only to provide food for people but also to assist them through replacement of lost income (Hay 1988). A special direct feeding programme was established in which health centres were given money, and then food was purchased locally from farmers or from co-operatives comprised of fishermen or herders. This strategy not only encouraged production but also injected more income into the rural areas (ibid.). Drought relief funds were used to build teachers' quarters and other facilities through Village Development Committees (VDCs); thus, the village had a longer-term stake in maintaining public works.

In 1984–5 a major remote area drought relief programme was established in Botswana in order to provide food and work for populations under stress in remote parts of the country. It should be noted that the western districts of Botswana, those with higher than average numbers of remote area dwellers, are the ones with higher levels of people who are nutritionally at risk. The selection criteria for those special remote area programmes involved determining those people who were beyond the reach of health facilities. By mid-1985 there were 20,000–21,000 remote area dwellers being fed. The food basket consisted of three basic commodities: (1) maize meal (12.5 kg per person per month), (2) beans (3.5 kg per person per month), and (3) vegetable oil (1 kg per person per month). This commodity basket was higher than that provided to nutritionally affected populations such as the vulnerable groups (under-5s, pregnant and lactating mothers) who were fed through the Institutional Feeding Programme (IFP) at schools and health posts in Botswana.

One of the more popular drought relief efforts in Botswana involved the destumping of agricultural fields. There was also a labour-based relief programme that served 170,000 people in rural Botswana, in which people were paid 2 *pula* (then about US$1/UK 30 p) per day to do such tasks as clearing roads and constructing storehouses and other facilities. Assistance was also provided in ploughing of fields. Tractors were made available by the Ministry of Agriculture to help people plough since in many cases their cattle were not in good enough condition to work, and their labour was divided because people were in towns seeking employment or at cattleposts taking care of cattle.

A major concern of the people of the northern Kalahari, besides the livestock disease situation, is that relating to human disease. Southern Africa was said to be the global epicentre of HIV/AIDS at the end of the millennium. Although the HIV/AIDS rate among the people of western Ngamiland appears to be much lower at present than is the case among other populations in Botswana, the infection rate is on the increase. Tuberculosis is also on the increase, linked in part to the spread of HIV/AIDS and other sexually transmitted diseases.

As has been documented so beautifully in the work of Katz, Biesele, and St Denis (1997), the Ju/'hoansi San of north-western Botswana engage frequently in healing rituals that are done through dance and going into trance. These healing practices are important means of curing the sick, alleviating stress, and bringing about positive changes in their communities.

Recently, the Ju/'hoansi and Herero of north-western Botswana have recommended that a 'community wellness programme' should be instituted that includes components to address problems of human and livestock health, nutrition, social conflict, and poverty. Having a culturally sensitive intervention programme that treats not only symptoms but also root causes of problems will go a long way towards ensuring that the people of the northern Kalahari will be able to live sustainably and in peace in the new millennium.

Acknowledgements

Support of the research upon which this paper is based was provided by grants from the US National Science Foundation, the Ford Foundation, and the US Agency for International Development. I wish to thank Deborah Sporton and David Thomas for inviting me to be a part of their symposium on Sustainable Livelihoods in Marginal African Environments held at Sheffield University, in Sheffield, England, 10–11 April 1997. Deborah Sporton and Chasca Twyman encouraged me to write up my research results and have provided me with useful insights on the situations facing Kalahari populations.

References

ALBERTSON, A. (1998), *Northern Botswana Veterinary Fences: Critical Ecological Impacts* (Ojai, Calif.: WILD Foundation).
BUTLER, D. (1996), 'Statistics Suggest BSE Now "Europe-wide"', *Nature*, 382: 4.

BARNES, J. (1994), 'Alternative Uses for Natural Resources in Botswana: Wildlife Utilization', in S. Brothers, J. Hermans, and D. Nteta (eds.), *Botswana in the 21st Century* (Gaborone: Botswana Society), 323–36.

BIESELE, MEGAN, and HITCHCOCK, ROBERT K. (2000), 'The Ju/'hoansi San under two states: Impacts of the South West African Administration and the Government of the Republic of Namibia', in Peter Schweitzer, Megan Biesele, and Robert Hitchcock (eds.), *Hunters and Gatherers in the Modern World: Conflict, Resistance, and Self-Determination* (Oxford and New York: Berghahn Books), 327–40.

CENTER FOR AGRICULTURAL SCIENCE AND TECHNOLOGY (1994), *FoodBorne Pathogens: Risks and Consequences*, CAST Report No. 122 (Ames, Ia.: Center for Agricultural Science and Technology).

CENTRAL STATISTICS OFFICE (1976), *The Rural Income Distribution Survey in Botswana 1974–5* (Gaborone: The Government Printer).

CHERFAS, J. (1990), 'Mad Cow Disease: Uncertainty Rules', *Science*, 249: 1492–3.

CHERNOCHOVSKY, D., LUCAS, R. E. B., and MUELLER, E. (1985), *The Household Economy of Rural Botswana: An African Case*, World Bank Working Papers, 15 (Washington DC: World Bank).

CONDY, J. B., and HEDGER, R. S. (1974), 'The survival of foot-and-mouth disease virus in African Buffalo with non-transference of infection to domestic cattle', *Research in Veterinary Science*, 16: 182–5.

CONDY, J. B., HERNIMAN, K. A. J., and HEDGER, R. S. (1969), 'Foot-and-mouth disease in wildlife in Rhodesia and other African territorie', *Journal of Comparative Pathology*, 79: 27–31.

DOBSON, A. (1995), 'The ecology and epidemiology of the rinderpest virus in Serengeti and Ngorongo conservation area', in A. R. E. Sinclair and P. Arcese (eds.), *Serengeti II: Dynamics, Management and Conservation of an Ecosystem* (Chicago: University of Chicago Press), 485–505.

DRAGER, N., PATTERSON, L., and BRETON, D. (1976), 'Immobilization of African Buffaloes (*Syncerus caffer caffer*) in large numbers for veterinary research', *East African Wildlife Journal*, 14: 1113–20.

FALCONER, J. (1971*a*), 'A history of the Bechuanaland Protectorate veterinary services', *Botswana Note and Records*, 3: 74–8.

——(1971*b*), 'Relationships between wild and domestic animals in the control of foot-and-mouth disease in Botswana', in Botswana Society (ed.), *Proceedings of the Conference on Sustained Production from Semi-Arid Areas* (Gaborone: Botswana Society), 153–6.

——(1972), 'The epizootiology and control of foot and mouth disease in Botswana', *Veterinary Record*, 91: 354–9.

——(1980), 'Disease situation report—foot and mouth disease—Botswana', *Bulletin de Office International des Epizootes*, 92: 655–7.

FALCONER, J., and CHILD, G. (1975), 'A survey of foot and mouth disease in wildlife in Botswana', *Mammalia*, 39/3: 518–20.

GIBBONS, W. J. (1970), 'Contagious pleuropneumonia', in W. J. Gibbons, E. J. Catcott, and J. F. Smithcors (eds.), *Bovine Medicine and Surgery: A Text and Reference Work* (Wheaton, Ill.: American Veterinary Publications), 179–81.

HARVEY, C., and LEWIS, S. R. JR. (1990), *Policy Choice and Development Performance in Botswana* (New York: St Martin's Press).

HAY, R. W. (1988), 'Famine incomes and employment: Has Botswana anything to teach Africa?' *World Development*, 6: 1112–25.

HEDGER, R. S. (1968), 'The isolation and characterization of foot-and-mouth disease in Botswana', *The Veterinary Record*, 91: 354–9.

——(1972), 'Foot-and-mouth disease and the African Buffalo (*Syncerus caffer*)', *Journal of Comparative Pathology*, 82: 19–28.

HINCHEY, M. T. (ed.) (1979), *Proceedings of the Symposium on Drought in Botswana* (Gaborone: Botswana Society).

HITCHCOCK, R. K. (1978), *Kalahari Cattle Posts: A Regional Study of Hunter-Gatherers, Pastoralists, and Agriculturalists in the Western Sandveld Region, Botswana* (Gaborone: Government Printer).

HITCHCOCK, R. K., and MASILO, R. R. B. (1995), *Subsistence Hunting and Resource Rights in Botswana: An Assessment of Special Game Licenses and their Impacts on Remote Area Dwellers and Wildlife Populations* (Gaborone: Natural Resources Management Project, Department of Wildlife and National Parks).

HITCHCOCK, R. K., YELLEN, J. E., GELBURD, D. J., OSBORN, A. J., and CROWELL, A. L. (1996), 'Subsistence Hunting and Resource Management among the Ju/'hoansi of Northwestern Botswana', *African Study Monographs*, 17/4: 153–220.

HOBBS, J. (1981), 'The environmental impact of veterinary cordon fences', *African Wildlife*, 35/6: 16–21.

HOLM, J. D., and MORGAN, R. G. (1985), 'Coping with drought in Botswana: An African success', *Journal of Modern African Studies*, 23: 462–82.

HUBBARD, M. (1986), *Agricultural Exports and Economic Growth: A Study of the Botswana Beef Industry* (London: Routledge & Kegan Paul).

HUNTINGFORD, T. G. H. (1990), 'Pleuropneumonia contagiosa', in *Diseases of Livestock*, 9th edn (New York: McGraw-Hill), 317–20.

KAPLAN, R. D. (1996), *The Ends of the Earth: A Journey to the Frontiers of Anarchy* (New York: Vintage Books).

KATZ, R., BIESELE, M., and St. DENIS, V. (1997), *Healing Makes Our Hearts Happy: Spirituality and Cultural Transformation among the Kalahari Ju/'hoansi* (Rochester, NY: Inner Traditions).

LAW, J. (1889), *The Lung Plague of Cattle: Contagious Bovine Pleuropneumonia* (New York: self-published).

MAUCO, P. M. (1978), *Income Effects of Foot and Mouth Disease to Cattle Owners: Information Note* (Gaborone: Ministry of Finance and Development Planning).

MERAFE, Y. (1978), *A Social Investigation into the Effects of Foot and Mouth Disease in Nokaneng, Tsau, Gweta, and Nata*, Report to the Rural Sociology Unit, Division of Planning and Statistics (Gaborone: Ministry of Agriculture).

MINISTRY OF AGRICULTURE (1996), *Contagious Bovine Pleuropneumonia in Botswana: Information Document* (Gaborone: Government of Botswana).

MINISTRY OF FINANCE AND DEVELOPMENT PLANNING (1977), *National Development Plan IV. 1976–1981* (Gaborone: Government Printer).

——(1985), *National Development Plan VI. 1985–91* (Gaborone: Government Printer).

——(1997), *National Development Plan 8: 1997/98–2002/03* (Gaborone: Government Printer).

MOTSHUBI, C. (1999), 'Experiences from Xai-Xai and Ukhwi', in L. Cassidy and M. Madzwamuse (eds.), *Community Mobilization in Community-Based Natural*

Resource Management in Botswana (Gaborone: IUCN Botswana & SNV Botswana), 45–7.

NTETA, D., HERMANS, J., and JESKOVA, P. (eds.) (1997), *Poverty and Plenty: The Botswana Experience* (Gaborone: Botswana Society).

NTSEANE, P. (1978), *Report on Foot and Mouth Social Investigation in Ngamiland and Central Districts*, Report to the Rural Sociology Unit, Division of Planning and Statistics (Gaborone: Ministry of Agriculture).

OWENS, M., and OWENS, D. (1981), 'Preliminary Final Report on the Central Kalahari Predator Research Project': Report to the Department of Wildlife, National Parks, and Tourism, Gaborone.

——(1984), *Cry of the Kalahari* (Boston: Houghton-Mifflin).

PEARCE, D., BARBIER, E., and MARKANDYA, A. (1990), 'Sustainable development in Botswana', in D. Pearce, E. Barbier, and A. Markandya (eds.), *Sustainable Development: Economics and Environment in the Third World* (Brookfield, Vt.: Edward Elgar), 150–67.

PEARCE, F. (1993), 'Botswana: Enclosing for Beef', *The Ecologist*, 23/1: 25–9.

PETERS, P. (1994), *Dividing the Commons: Politics, Policy and Culture in Botswana* (Charlottesville and London: University Press of Virginia).

PFOTENHAUER, L. (ed.) (1991), *Tourism in Botswana: Proceedings of a Symposium Held in Gaborone, Botswana, 15–19 October, 1990* (Gaborone: Botswana Society).

PRUSINER, S. B. (1997), 'Prion Diseases and the BSE Crisis', *Science*, 278: 245–51.

REPUBLIC OF BOTSWANA (1975), *National Policy on Tribal Grazing Land*, Government Paper No. 2 of 1975 (Gaborone: Government Printer).

——(1986), *Wildlife Conservation Policy* (Gaborone: Government Printer).

——(1998), *Community-Based Natural Resources Management Policy* (*Draft*) (Gaborone: Government of Botswana).

RUIGROK, E. A. (1997), *Revised Project Memorandum, CBNRMP /Xai/Xai: Project Review 1994–1997 and Project Formulation 1997–2000* (Gaborone: SNV (Netherlands Development Organization)).

SALKIN, J. S., MPABANGA, D., COWAN, D., SELWE, J., and WRIGHT, M. (eds.) (1997), *Aspects of the Botswana Economy: Selected Papers* (Oxford: James Currey, and Gaborone: Lentswe La Lesedi Publishers).

SANDFORD, S. (1977), *Dealing with Drought and Livestock in Botswana* (Gaborone: Government Printer).

——(1979), 'Towards a definition of drought', in M. T. Hinchey (ed.), *Proceedings of the Symposium on Drought in Botswana* (Gaborone: Botswana Society), 33–40.

SCHAPERA, I. (1943), *Native Land Tenure in the Bechuanaland Protectorate* (Alice, South Africa: Lovedale).

SILBERBAUER, G. S. (1981), *Hunter and Habitat in the Central Kalahari Desert* (Cambridge: Cambridge University Press).

SMIT, P., and KAPPE, S. (1992), 'Socio-Economic Baseline Survey of Ngamiland's Remote Zone', Report to the Western Communal Remote Zone of the North West District Land Use Planning Unit and to the Ministry of Local Government, Lands, and Housing (Gaborone).

VAN ONSELEN, C. (1972), 'Reactions to Rinderpest in Southern Africa 1896–97', *Journal of African History*, 13/3: 473–88.

WIESSNER, P. W. (1977), 'Hxaro: A Regional System of Reciprocity for Reducing Risk among the !kung San', Ph.D. Dissertation, University of Michigan, Ann Arbor, Michigan.

WILHITE, D. A., and GLANTZ, M. H. (1987), 'Understanding the drought phenomenon: The role of definitions', in D. A. Wilhite, W. E. Easterling, and D. A. Wood (eds.), *Planning for Drought: Toward a Reduction of Societal Vulnerability* (Boulder, Col.: Westview), 11–27.

9

Drought as a Revelatory Crisis Revisited: An Exploration of Shifting Entitlements and Hierarchies

Jacqueline S. Solway

Introduction

In his analysis of the domestic mode of production Sahlins (1972: 124, 143) refers to a revelatory crisis as one of the few occasions outside of an 'act of ethnographic will' that exposes to the observer the central contradiction in that mode of production. The drought in southern Africa that occurred in Botswana in varying waves of intensity between 1979 and 1987 constituted such a revelatory crisis.[1] It precipitated the interruption of socio-economic patterns to a degree sufficient to lay bare contradictions in the existing order that had been latent or contained prior to the drought. Structural contradictions such as those between household and kindred, between individualized and communal property claims, between production for market and production for subsistence, between the state's vision of 'rational' peasant production and the realities of daily economic life in the rural areas, and those between classes were revealed to varying degrees and in different ways both to the outside observer and to the participants themselves. Moreover, the drought also exposed the extent to which conditions in the rural areas had deteriorated. In doing so it revealed a 'crisis of social reproduction' (cf. Watts 1987: 207); the crisis at hand was not simply one of external origin but was systemic and indicative of long-term change.

This chapter is an abbreviated version of a paper that appeared in *Development and Change*, 25 (1994), 471–95 and is reproduced with permission from Blackwell Publishers, Copyright 1994. This revised version includes a postscript to update the material.

[1] Sahlins draws upon the work of R. Firth on Tikopia for this analysis. The specific revelatory crisis was a famine.

Things were not going to 'get back to normal' when the rains resumed. And indeed, they have not.

Southern Africa is a drought-prone region. Droughts occur with alarming frequency in Botswana, on average every seven years. Therefore, one can assume that a cyclical pattern must exist. However, I argue that this drought broke the cyclical pattern and fundamental as opposed to superficial changes became entrenched. It occurred at a critical historical juncture in Botswana; commercialization and class formation had been proceeding at a rapid pace[2] and Botswana's economy, newly rich from diamond revenues, was growing dramatically. The drought of the 1980s was thus a 'watershed' event. Changes in meaning and practice reached a point where pre-drought production patterns, in all likelihood, would not return. In addition, the state initiated actions to ameliorate the effects of drought which, in many respects, hastened the trajectory of change and, at the same time, laid the basis for its own expanded presence. The drought was thus an opportunity and provided a point of entry for the state to insert itself in the lives of citizens in new and expanded ways (cf. Ferguson 1990).

However, in a paradoxical fashion, at the same time as a crisis such as a drought reveals and exposes contradictions and deteriorating conditions, it also allows them to be concealed and mystified. It is precisely because drought is discursively constructed as a crisis and therefore believed to be temporary, unusual, externally wrought, and severe in nature that it allows systemic socio-economic problems to be concealed. Drought is a perfect scapegoat; all social and economic dislocation and suffering can be attributed to the drought and underlying problems can be left unacknowledged and therefore, unconfronted. This is what has happened in Botswana when both state officials and many citizens alike argued that the problems surrounding the drought would go away with the return of the rains. This reluctance to admit to a fundamental crisis of social reproduction contributed to the formulation of drought relief measures which, in some respects, were palliative in the short term but in the long term have served to exacerbate socio-economic inequalities.

The third point I wish to make with regards to drought is that it offers a critical occasion when conventional routine is sufficiently disrupted that actors are given licence to innovate with social and moral ideological and behavioural codes. It is a time of experimentation; when taboos can be violated, moral codes flouted, and tradition invented (Hobsbawm and Ranger 1983). Something that may have been unacceptable before is possible during a drought. Thus drought is a perfect venue for agency and change and therefore offers a lens through which to view the dialectic between structure and agency. The opportunities actors take when certain structural fault lines are exposed are not random nor are they entirely predictable. The process of how such change occurs and what aspects of a system are emphasized, elaborated,

---

[2] This is particularly true in the Kalahari region where wealthy cattle ranchers began pursuing commercial production on a greatly expanded scale after a borehole loan scheme in the 1960s and the opening up of livestock marketing co-operatives in the 1970s.

or altered are analysed here. Thus, change is revealed as neither the simple unfolding of the logic of a preordained system (such as capitalism) nor as entirely the result of individual agency.

This chapter draws upon extended case material from Central Botswana based on fieldwork covering a period of seventeen years. It is argued that a decline in the capacity to produce amongst Botswana farming households is related to commercialization, changing conceptions of property, and state intervention. Sen's method of entitlement analysis is utilized to examine social process in the realm of property relations in which narrow exchange rights have been increasingly granted legitimacy over wider use rights. The incomplete shift in entitlements from locally based to state-based sources is also examined. In the final section dependency relations are considered and questions relating to various forms of patronage are posed.

Drought and Agriculture in Botswana: The National Context

In post-independence Africa, drought[3] has been reputed to be a leading cause of declining production and great human suffering.[4] Yet the manifestations of drought and people's experience of drought conditions are mediated at all levels by non-meteorological factors. Moreover, within rural communities and even within households the effects of drought are not evenly borne. When the statistics are disaggregated it often comes to light that neither is declining production and/or food shortage generalized between households nor is suffering equally distributed within households; certain members, particularly women, children, and retainers, are often the first to suffer.[5] Although for most (but not all) of the years between 1979 and 1987 Botswana's rainfall levels were sufficiently below average to rank her amongst the five most drought-stricken countries in Africa (Holm and Morgan 1985) the government managed to keep famine at bay. Massive drought relief measures including food relief, labour-based relief, water development, and agropastoral subsidies ensured that no one starved.[6] Every rural resident received, in one form or several, the benefits of drought relief. Botswana's capacity to

---

[3] For the purposes of this chapter, drought is defined simply as significantly lower than average rainfall levels.

[4] See e.g. Bush (1988); Dreze and Sen (eds.) (1990); Franke and Chasin (1980); Glantz (ed.) (1987); Lofchie and Commins (1982); Shipton (1990); Swift (1993).

[5] See e.g. Berry (1984); Vaughan (1987); Watts(1987); Wylie (1989).

[6] Relief measures escalated during the latter period of the drought; they included the development of an early warning system for the detection of malnutrition, agro-pastoral subsidies, and an emphasis on labour-based employment generation relief (Dreze 1990: 153–5). In 1985 the equivalent of 26,413 full-time jobs were created by the drought relief programme. In the same year the formal sector (excluding drought relief) accounted for 105,000 full-time job equivalents (Valentine 1993: 120). According to the 1990 Botswana Development Plan (21), between 1982 and 1990 approximately US$200 m. was spent on drought relief. In the mid-1980s, 380,000 people (over a third of the population) were *direct* recipients of food relief (Republic of Botswana 1989, emphasis added).

generate relief derived from her own relative wealth,[7] from the receipt of foreign aid, and from the fact that the delivery of relief was handled efficiently and with remarkably little corruption.

This was a stunning achievement and has been cited as an exemplary case of famine prevention (Dreze 1990: 151–8; Hay 1988; Valentine 1993). However, although starvation was averted, the drought relief measures have served to accentuate socio-economic inequalities and have quite possibly undermined the capacity of the majority of rural peoples to produce their own subsistence and to support themselves.

On a national level almost 40 per cent more hectarage (408,000 ha) was under cultivation by the end of the drought in 1988–9, than was under cultivation in 1977–8 (260,000 ha) (Republic of Botswana 1991: ii. 10). Given that farmers were literally paid to plough by the government from 1985 to 1990 the increase in hectarage is not surprising. But one must question seriously whether such figures will be sustainable once the subsidies are withdrawn and farmers must fund their own agricultural initiatives. In addition, the figure for the increase in hectarage must be tempered by information regarding who is doing the farming. Mean farm sizes grew in the 1980s but the increased hectarage under cultivation was not equally distributed (ibid. ii. 4). Drought relief agricultural subsidies enabled some farmers, particularly the larger and wealthier ones, to enlarge their farms dramatically while, at the same time, many smaller, particularly resource-poor farmers, were unable to take full advantage of the programmes (discussed in more detail below). Moreover, many poorer farmers found that the arrangements they had relied upon prior to the drought to gain access to agricultural fields and resources were no longer as readily available.

The fact that for years after the drought's end the government was still at pains to withdraw relief again indicates that the crisis exposed by the drought was one the roots of which go deeper than the recent rainfall shortages. After the drought ended in 1987 certain relief measures were simply renamed drought recovery, some have been incorporated into Botswana's expanding and ongoing social welfare programme,[8] and some were withdrawn in 1990.

---

[7] By most standards, Botswana is an African economic miracle. The World Bank *Report* (1992: 219) reveals that Botswana had the fastest-growing economy in the world between 1965 and 1990; its growth rate had averaged 8.4% p.a. during this period (South Korea, the second fastest-growing economy, averaged a growth rate of 7.1%). As a result, according to the United Nations, Botswana has gone from being amongst the twenty-five poorest nations in the world at independence in 1966 (per capita income was approximately US$80 (Colclough and McCarthy 1980) to being one of Africa's wealthiest in the 1990s (in 1991/2 per capita income was approx US$2,850 (Republic of Botswana 1993: 5). However, these stunning figures should not disguise the existence of marked discrepancies in wealth and deepening inequalities within the country (Good 1992).

[8] For instance, between 1984 and 1988 the number of registered destitutes receiving a monthly government allowance doubled. By 1990 there were as many official destitutes as there were employees in the mining sector which, in 1990, accounted for 80% of Botswana's export earnings (Good 1992: 90). Despite the return of rain, subsidies continued for years under the guise of drought-related programmes.

In keeping with a time-honoured anthropological tradition of illuminating the general with the particular I will employ a specific example to illustrate the general processes outlined above.

The Local Context

My specific case material derives from three periods of fieldwork in central Botswana.[9] I worked in the Kalahari region of the western Kweneng district, a semi-arid area receiving approximately 300–50 mm of rainfall a year. The predominant local 'ethnic' group, both politically and numerically, are the Kgalagadi (cf. Kuper 1966, 1970a, 1970b). The Sarwa (San or Bushmen) who predate the Kgalagadi in the area have a history of hunting and gathering but now largely constitute a casual labour force for the Kgalagadi. The Sarwa are the largest per capita recipients of food relief and have increasingly come to rely on this as their basic food.

The local economy is based on pastoralism (most importantly, cattle), supplemented by rain-fed agriculture (beans, maize, melons, and sweet-reed are grown, but principally sorghum). Livestock are valued for the meat, milk, dung, and draught-power they provide; they are valued for purposes of local exchange, both as gifts and as payments; they are valued for the social and spiritual sense of well-being they afford their owner; for the role they play in mediating relations between people, and between people and the *badimo* (ancestors) (cf. Bonte 1975, 1981: 40–1). Cattle are used in curing ceremonies and are sacrificed at rituals; as such they offer a direct link between the material–social domain of people and things and the spiritual domain of gods and power. And, of course, cattle are valued for the price they obtain at the market. Botswana maintains favourable trade arrangements with the EU who imports her beef at inflated prices. The use of cattle for subsistence purposes undermines their market value, thus the two types of production are not easily compatible. Since all cattle-owners are simultaneously involved to varying degrees in both types of production, they are constantly weighing the relative costs and benefits of pursuing the different production strategies.[10]

In addition, virtually all Kalahari residents are dependent, directly and indirectly, upon income derived from wage labour. Until the late 1970s most men migrated at several points in their adult life to the South African mines (in 1978, 84% of Kgalagadi men I interviewed had been to the mines). By 1980 South Africa had cut back its recruitment of foreign miners, in effect withdrawing an employment opportunity upon which rural residents had

---

[9] Twenty months in 1977–9, eight months in 1985–6, and very briefly in 1990.

[10] Subsistence-oriented and commercial or market-oriented production rationales are best viewed as 'ideal types'; neither exist in pure form on the ground but represent poles in a continuum upon which existing situations can be placed.

come to rely. Extremely few wage opportunities existed for women at that time. New formal sector employment opportunities have become available in Botswana since 1980.[11] However, the majority of new jobs have been for skilled and educated individuals thereby giving advantage to better-off households who had previously invested in their children's education.[12] Thus the trickle-down effect of transfers from formal sector employment are not as equally distributed within rural communities (cf. Good 1992: 92) as South African mine wages were in previous decades. Examples such as this point to the ironies of development.

The Kalahari is rich cattle country, covered with good pasturage and few, if any, poisonous plants for cattle. The limiting factor on pastoral production in the region had been water, but since the 1960s boreholes have continued to produce new sources. In the 1980s drought, for the first time, according to my informants, pasturage became a limiting factor in this part of the Kalahari. Agriculture is risky because of erratic rainfall but twenty years of data indicate that for farmers well prepared to plough, good harvests outnumber bad ones.[13]

Social Differentiation

In order to understand the role and consequences of drought relief measures in the region it is necessary to summarize recent changes in patterns of social differentiation in the Kalahari (see Solway 1979; 1987 for a fuller discussion of these issues). Tswana–Kgalagadi structures have long been ranked and disparities in political, social, and economic status between households have existed for generations. However, in recent decades it has become apparent that the nature of social differentiation has been transforming; since the 1960s, in particular, opportunities have become available for significant capital investment in productive resources, and this along with explicit state sanction for commercial development has resulted in the formation of a distinct group of accumulators (cf. Peters 1984).

[11] Formal-sector employment has grown at a rate of 9% p.a. since 1966 while the population has grown at a rate of 3.4% in the same period (Republic of Botswana 1991: ii. 19). Still, unemployment is a serious problem.

[12] Only in the 1980s did public education become free.

[13] This refers to data I collected in one village based on personal observation and the memories of many informants. The notion of 'good' versus 'poor' yield is a subjective one and is derived from my informants. A good yield does not imply self-sufficiency in grain; rural Batswana have long been incorporated into a money economy and prefer to vary their diets with purchased carbohydrates. A good yield ideally gives a household enough sorghum for its own defined nutritional needs and a sufficient surplus from which to sell, make gifts, pay agricultural labourers, brew beer for ritual events, and display hospitality. The largest harvest of a village household in 1978 (the last year before tractors were used and yields began to vary to a much greater degree) consisted of approximately 2,940 kg of sorghum, 665 kg of maize, and 105 kg of beans. The household sold approximately 25% of the harvest, used about the same in payment for labourers, gave less away as gifts, and reserved the rest for household consumption.

Although the process is at best uneven, capitalist development and class formation are occurring. Amongst the rural élite there has been a shift of emphasis from wealth in people to wealth in things and an increasing trend towards privatization and consolidation of economic resources. This changing emphasis has contributed to a productive strategy on the part of the emerging commercial élite which has led them to withdraw, to a certain extent, their economic resources from the pool available for communal use, a process that is affecting the pattern of social reproduction. Thus poorer members of the community experience greater difficulties in gaining access to productive resources both for purposes of use and for accumulation. In this manner discrepancies in productive potential are growing; accumulation is creating a parallel process of dispossession.

The key resource and marker of wealth in the Kalahari has been cattle; herd size in the region ranges from zero to approximately 1,000 with most cattle-owning households having less than 40 head.[14] The process of commercialization leads many wealthier households to emphasize the market value of their livestock over local use and exchange value. Thus commercially oriented herders are less likely to milk their cows, to plough with them, to exchange them as gifts, sacrifice them at rituals, or to let others use them for these purposes. What is significant is that such decisions by the commercial élite have ramifications far beyond their households since the majority of villagers have depended upon access to the resources of the wealthier (to be illustrated below with regards to agriculture).

The current pattern of accumulation reverses previous patterns. For the first two-thirds of the twentieth century accumulation had a dual dynamic of both concentration and dispersion of property. For instance, at the turn of the century the Kgalagadi were emerging from a period of political subjugation in which their capacity to accumulate property was circumscribed by the powerful Tswana chiefdom to their east. Initially, in the early part of the century, the political élite accumulated livestock; by 1977 through local means of redistribution and through migrant labourers' wages only 8.6 per cent of Kgalagadi households in the community where I conducted research owned no cattle; in 1986 that rate had doubled to just over 17 per cent. National figures are consistent with my own. According to the 1991 National Development Plan farming households (thus excluding urban-based and poor rural households who do not farm) with no cattle increased from 28 per cent to 38 per cent between 1980 and 1988 (Republic of Botswana 1991: ii. 4).

Changes in the productive process are part of a larger complex of changes in which the whole pattern of social reproduction is implicated. For example, ideologies based on Western Christian individualism compete with those

---

[14] Increasingly and particularly since the 1980s new forms of wealth and status have become important. These include vehicles, 'modern' homes, and, especially, valued formal-sector employment within Botswana.

grounded in the ancestors which affirm principles of mutual responsibility in which one's own well-being is inseparable from that of one's kin. There has been experimentation on the part of some wealthy households with 'new' non-syncretic Christian sects; these illustrate attempts by the élite to find cosmologies more consistent with their emerging worldviews.

The local élite do harbour some ambivalence towards their declining role as social and economic benefactors. Business is easier to manage without attending to the diffuse needs of others, but, at the same time, benefactors or patrons lose labour and social and political support, and become subject to retribution, the 'weapons of the weak' (Scott 1985). None the less, a diminishing of diffuse organic links between rich and poor in the region is a discernible trend.

Agriculture

At this point I wish to focus upon changes in agricultural production between the late 1970s and late 1980s in order to illustrate the growing discrepancies in productive potential that were thrown into relief by the drought and the manner in which agricultural subsidies, most of which were part of the drought relief measures, provided differential benefits to the rural populace depending on their position within the local political economy. In addition, it will be shown that the agricultural subsidies were based on certain ideological notions that in themselves contributed to reformulating rural relations and realities with the effect of further reinforcing growing disparities.

I will present this data by comparing participation in agricultural production during the two seasons for which I collected survey data, 1977–8 and 1985–6, and for 1990, a year for which I have less complete data. In 1977–8 approximately 90 per cent of Kgalagadi households (n = 70) in the village where I based my fieldwork ploughed.[15] In 1985–6 the number of households ploughing was reduced to about half that figure. This statistic is rendered all the more striking by the fact that in 1985–6 there were three government-sponsored programmes offering substantial subsidies for agricultural production and in the late 1970s there were none. The rains that herald the ploughing season came about two weeks later in 1985 than they had in 1977 and people claimed to be demoralized because of a series of years of poor rainfall. However, neither of these factors can entirely account for the discrepancy, especially since 1978 followed a year of poor harvests and the agricultural subsidies of the mid-1980s should have compensated for some of the demoralization. The later rains of 1985 were not insignificant in terms of limiting numbers of people ploughing but they cannot begin to account for

---

[15] Sarwa households, which comprise less than 20% of the total, rarely pursue independent agro-pastoral production.

the discrepancy between the years. Late (but not too late) rains usually account more for fewer hectares under cultivation than for fewer households ploughing. In the pre-agricultural subsidy years households often reduced risk by ploughing less when the rains were late. In addition, rates of field sharing increased in such a year. Many poorer households, often in a queue for oxen or wishing to minimize risk in a questionable year reduced their input into field preparation and, instead, attempted to obtain a small plot in the fields of another household that had better access to draught-power and fenced and cleared fields. Participation in agricultural production declined almost steadily in the first half of the 1980s. It rose again in the second half of the decade (although not to 1978 levels) but virtually all the households ploughing were doing so at government expense.

'Explanations' of Declining Participation in Production

Before discussing explanations provided for declining production, it is important to distinguish declining levels of production in the form of output or overall yield from declining levels of production in terms of the capacity to produce amongst individual farmers. These two aspects of production may vary independently of each other but are often conflated in discourse on the effects of drought. In a situation of rain-fed arable production the impact of drought on yields is, to a great degree, a technical matter. In dryland farming measures can be taken to offset the effects of drought but insufficient or poorly timed rain can still reduce or eliminate a crop. However, the capacity to produce refers to series of relationships between producers which regulates their access to productive resources. If fewer farmers have the means to produce then aggregate harvest may be diminished. But, of course, this may not necessarily be the case if the means of agricultural production become concentrated in fewer hands—fewer large farmers could produce more food than many small farmers.

In Botswana it is clear that both the aggregate harvest and the numbers of people who could produce were diminished at the height of the drought. The two are related but, as noted above, not in a simple linear or consistent fashion. For the purposes of this chapter, I am primarily concerned with changing patterns of agricultural production in terms of the capacity to produce.

In the latter half of the 1980s, drought was cited in almost every instance by local people as the explanation for declining participation in agricultural production and for declining yields. Official national discourse also implicated the drought as the cause for both. However, the more I investigated the situation the more this answer seemed inadequate. National statistics showing lower than average rainfall levels for several years could not be contested but they only tell a partial story; local and yearly variation was significant. In the village where I worked most households claimed they had not reaped a harvest in years because of the drought, yet some households reported years of bumper harvests. One even proudly showed me the 'sound system' which was purchased from the profits from selling one year's surplus harvest. In addition, the discrepancy in harvest sizes had grown significantly. In 1978 while the largest harvests of the principle crop, sorghum, were approximately 40 bags; by the mid-1980s the largest harvests consisted of approximately 140 bags. The vast majority of households continued to harvest well under 10 bags. Such patterns of 'feast during famine' certainly resonate with Sen's observation that, 'Indeed, in many famines complaints have been heard that, while famine was raging, food was being *exported* from the famine-stricken country or region' (1982: 161 emphasis in original).

Certainly the causes of these stark discrepancies are complex and cannot be explained by any single factor. Insufficient rainfall played a role but that role has been inconsistent; some households increased production while for most production declined. To understand changes in productive capacity it is necessary to move from technical factors and to examine socio-economic relations.

It is useful at this point to evoke Sen's method of entitlement analysis. Sen (1982) locates the source of scarcity, food in this instance, in the realm of social relations which govern people's socially recognized access (entitlement) to a desired good rather than in the absolute supply of the good. Entitlement refers to a person's legitimately acknowledged right to a certain thing in a specific society. Entitlement relations are linked, in Sen's words (ibid. 2), into chains in which one set of entitlements legitimizes another set. Following Sen's typology, what occurred in Botswana was that state-based social security entitlements expanded and averted potential famine. Farmers whose harvests failed or those that were unable to plough could command food directly through government programmes or indirectly through programmes that offered employment or provided agricultural subsidies to individuals and households. These entitlements formed links in an entitlement chain that could eventually lead to entitlement to food. Sen (ibid. 7) notes that in countries such as Britain and the United States where unemployment is high there would be widespread starvation 'but for the social security arrangements' which grant people entitlement to a number of basic provisions. In Botswana the expansion of social security entitlements compensated for and, at the same time, masked the countervailing process in the realm of locally based entitlements in which there was a contraction. This will be illustrated when I discuss agricultural arrangements in more detail.

In the 1970s the entitlements that operated that permitted such high levels of agricultural production included those which granted access to the three most important resources (besides rainfall) in Kalahari agricultural production draught-power, land, and labour. Of these three, draught-power has always been the most critical for Kalahari residents.

Draught-Power and Property Relations

Apart from the relations that existed around reciprocal access to water sources, particularly those for watering livestock, the relations which existed around reciprocal access to draught-power were the most significant in binding the community in an ongoing network of rights and obligations during the 1970s. It was access to these resources that people often gave as reason for moving from one village to another and it was the relations surrounding access to these resources that often formed the basis upon which other sets of relations were built.

Prior to the 1979–80 agricultural season all draught-power in the region where I conducted fieldwork was animal, predominantly oxen. Approximately half the households ploughing (90% of Kgalagadi households) in 1978 owned sufficient oxen to form their own spans of six; the other half utilized one or more of the many formal and informal relations that facilitated access to draught-power. At the time none of these relations were commodified in the sense that money changed hands. However, they frequently entailed labour exchange as well as more diffuse exchanges not directly tied to the oxen exchange but that were part of the wider relations between the parties.

Local conceptions of property facilitate the reciprocal use of resources. Rights to property are simultaneously individuated and dispersed (cf. Galaty 1981: 10), i.e. virtually all items of productive property can be identified with an individual but at the same time a larger group maintains rights to or, in other words, has entitlement to the 'family estate' (Gray and Gulliver 1964; cf. Carstens 1983: 60). Communal property claims rest, to a great extent, upon the notion that most property derives ultimately from the ancestors and was created and sustained through the labour of past family members; their descendants can claim the fruits of this labour and the patrimony of their forefathers.[16] In most Kgalagadi communities Kgalagadi residents are linked through dense agnatic and affinal links so that virtually everyone shares the same ancestors. However, inheritance patterns are such that each item of property, e.g. each calf, is designated as belonging to one individual although the wider claims are implicit. The rights to dispose of or to exchange an item of property such as a cow typically rest with that individual (if he or she is not a minor) *and* a small circle of kin.[17] Such decisions are conventionally made

[16] This system, of course, denies the very significant contribution that Sarwa labour has had in building Kgalagadi wealth. See Solway (1979; 1987), Solway and Lee (1990), Hitchcock (1987), and Wilmsen (1989) for a discussion of the ways in which ethnicity operates in Kgalagadi property ideologies to exclude the Sarwa from claims to productive resources.

[17] I have deliberately left the categories of small versus wide circles of kin vague in order to emphasize the relational quality of rights and obligations within a field of relations. This also facilitates comparison with other cases along the lines outlined in the paper. Amongst all peoples there are cultural assumptions regarding the differences between kin and the types of access they have to certain resources. Amongst the Kgalagadi these differences are influenced by considerations of the past, the present, and the future with agnatic descent being the most important constant variable. Property

through consultation amongst a group of kin; this group is smaller than those who can exercise claims to the use value of the property. Thus claims to the use value of a particular item tend to be wider than claims to the exchange value.

The manner in which claims can be articulated and the network of relations they can implicate is illustrated by the following example. In 1986 a relatively poor widow told me she had recently ploughed her fields with her 'father's oxen'. This surprised me, as not only had her father been dead for eight years but, being extremely old (he was over 100 when he died), he had divided his cattle and the authority and responsibility over them amongst his children at least a decade before his death—close to twenty years before the woman ploughed with her 'father's oxen'. Upon further questioning it became clear that the cattle she ploughed with were from her brother's kraal, had the brother's brand, had been his responsibility for decades, and were largely considered the brother's cattle. Yet, at the same time, the deceased father's association with the cattle remained alive, justifying the woman's claim to them. In this case the woman's phrasing (that the cattle belonged to her father) already took her well into the process of exercising her kin-based entitlement to the cattle.

The process of commodification leads to an emphasis, particularly amongst the more commercially oriented households, on the private end of the property continuum. However, such an emphasis does not necessarily or easily result in the relinquishing of the communal end. Similarly, the expansion of commodification also grants greater, but not complete, legitimacy to the rights of the narrow group who can command exchange value over the wider group that can command use value. Thus while one person's claim may rule out the viability of another's, it does not preclude the other claim from being made nor does it negate its legitimacy. Such a system invites redefinition, negotiation, and a constant series of claims and counter-claims amongst kin (see Peters 1992 for a discussion of similar processes operating in the redefinition of rights to boreholes and grazing land).

Depending upon their location in the local political-economic structure, farmers have different and, at times, competing interests and interpretations as to the extent of legitimate claims over property. But whatever their interests, local property conceptions still render a variety of often contradictory claims legitimate and denying them requires a rationalization that is rarely universally accepted.

---

devolves agnatically to all descendants and the closer the relation along agnatic lines, generally the stronger the claim. However, sisters' claims are compromised by the fact that ideally their children belong to another agnatic unit and the locus of their rights and responsibilities lies elsewhere. However, given the fact that discrete descent groups do not form amongst the Kgalagadi because marriage is permitted amongst patrilateral and matrilateral cousins, because many women have children outside marriage, and because of the myriad contingencies of everyday life, it is not always practical to observe ideal rules based on agnatic ideology and descent.

A locally recognized crisis such as a drought offers an opportunity for a redefinition of the range and priority of property relations and claims. Although people suffer differentially during a drought, they all experience and acknowledge its negative effects. Thus it provides one of the most easily accepted rationales and, indeed, to some extent, a licence and moral pretext for the act of denying communal claims, an act that may have been thought to be antisocial prior to the drought. In some instances communal claims may be difficult to honour because of the constraints imposed by late or limited rains. For instance, a common situation exists when a man usually ploughs for his widowed mother's household or unit and his unmarried sister's in addition to his own. When the rains arrive late, and the ploughing season is truncated, he can only plough for one or two. A decision must be made as to which unit's entitlements are the most attenuated—most likely it would be the unmarried sister and she and her children would suffer first. However, not all denied claims are as directly justified by climatic conditions as will be seen later in the chapter in cases where certain farmers withdraw their oxen from ploughing altogether and turn to tractors.

In the least formal arrangements oxen are simply shared. Between closely related households exchanges occur frequently; a draught power exchange would simply be embedded within an ongoing exchange cycle and not specifically linked to any direct reciprocal exchange. More formal arrangements for gaining access to draught-power include *mafisa*, a system of loan cattle whereby a herder takes one or several of another's cattle, cares for and uses the beasts for an unspecified number of years, receives a female calf as compensation, and then eventually returns the animals. Households short of oxen, but not of male labour, often ask a wealthier household to loan them one or more oxen for the ploughing season. Typically young oxen are loaned and the borrower is responsible for the extremely time-consuming and difficult process of training them. After one or two ploughing seasons the owner has returned trained oxen. Another common strategy, particularly for households short of male labour, is to help another plough and then take the oxen, and occasionally the plough and labour, to their own fields. This option is the least likely to be utilized in a season of late rains. All these methods are most rational when subsistence livestock production is emphasized over commercial.

Tractors have rapidly replaced oxen as the preferred form of draught-power. By 1990, in the village where I worked, three households owned tractors and several from elsewhere were available for hire. Tractors use little labour and can cover large areas quickly. Since the mid-1980s the government has subsidized tractor rental and, according to a senior government official with whom I spoke in 1990, the net effect of this policy has been that many tractor-owners have become very wealthy. One local tractor-owner, for instance, ploughed for twenty-three people in addition to himself in 1990, all on government subsidies. However, the use of tractors has its consequences. The long-term ecological consequences of tractor use are no doubt significant, and the socio-economic consequences, particularly in a situation where the vast majority of farmers cannot, on their own, afford tractors or their rental, are many.

Agriculture is risky in the Kalahari; ploughing is no guarantee of harvest and crops can scorch at any point in the season. The greater the investment, such as that involved in tractor use, the greater the risk. In the last several years the government has assumed the risks in tractor agriculture by bearing its monetary costs, but this is unlikely to be sustainable.

Farmers choose tractors over oxen because ploughing with oxen diminishes their market value. Prime, young, unworked oxen command high prices at the abattoir. Already, by the early 1980s, before government tractor rental subsidies became established, commercially oriented households were attempting to limit their oxen's use. Some were beginning to rent tractors and many were keeping fewer mature oxen. Therefore when others asked to use their oxen they could truthfully say that they had insufficient to loan. This is an example where wide claims to use value can be precluded by narrower claims to exchange value. Herd composition figures reveal the change: for example, in a sample of moderately large herds—50 to 150 head (n = 10)—herd composition figures from 1978 indicate 13 per cent of work oxen in the herds, while in 1985 the figure for the *same* herds was approximately 6 per cent. Similarly, in 1978 approximately 50 per cent of farming households still had sufficient oxen to compose a ploughing span as opposed to approximately 10 per cent in 1985. Because of the drought some cattle had been sold to meet food shortfalls, and pasture and water shortages had weakened the animals. But herd composition figures did not alter in similar fashion in the 1960s when an equally serious and long drought occurred. And it is important to note that the trend in tractor use and withdrawal of oxen from ploughing *preceded* the drought; the drought certainly hastened the process and it also provided a perfect rationale for farmers to legitimately dispose of their oxen. The fact that in 1990, three years after the drought officially ended, there were proportionately fewer oxen than at the height of the drought is testimony to the fact that other factors than the drought have contributed to the decline in numbers of draught animals. In 1990 I was told (although I was unable to conduct a systematic investigation) that only three households still kept plough oxen.[18] As I have emphasized, such a reduction in oxen affects the ploughing potential of the community at large and not just the individual households who choose to sell their oxen.

Tractors have not been entirely assimilated into the property concepts that surround other productive resources such as livestock. The link between tractors and the patrimonial estate is not absent but is less clear, direct, and, to many, less compelling. It can be argued, and some make this argument, that the patrimony was used to purchase the tractor: patrimonial wealth was

---

[18] Milking practices have followed parallel patterns. In the late 1970s all cattle-owners milked and milk was a dietary staple for ill members of the community. Powdered and UHT milk, unavailable locally in 1978, lined the shop shelves in 1985.

simply converted into cash. Cash also has not been entirely assimilated into existing property relations; its association with individuals is much stronger than other forms of property. And cash, unlike livestock, is not visible. This is a point again where the legitimacy of narrow rights to exchange value take precedence over wider communal rights in an increasingly commodified system. In addition, tractors, for a variety of reasons, do not lend themselves to the same sorts of reciprocal arrangements as cattle do. Their high initial cost, their maintenance and running costs, and the fact that they do not reproduce themselves has contributed to their lack of integration into existing entitlement patterns.

The majority of oxen-sharing arrangements involve some sort of labour exchange and many are premised upon asymmetrical relations. Thus they are not great equalizers; but they are levelling mechanisms that function to keep the gap between rich and poor narrower than otherwise might be the case. They are relations of entitlement (or endowment in the sense that they form the basis of a person's entitlement) that permit semi-independent production on the part of the poorer majority. As such they are consistent with the Kgalagadi cultural logic of self-construction (cf. Alverson 1978: 135; J. L. Comaroff 1982: 109; Comaroff and Comaroff, 1987), the ethic of doing for oneself that the Kgalagadi contrast with other forms of economic activity, particularly wage labour, that do little to contribute to an individual's 'valorization' claims to legitimacy and social standing (Ortner 1989). Such entitlements, unlike state-based social security entitlements, contribute to constituting the individual with dignity as defined within Kgalagadi cultural models and hold out the possibility that any individual will eventually assume and achieve a position in life where they will be able to grant entitlement to others and the political and economic benefits that entails.

Land, Labour, and Agricultural Subsidies

Land and labour, the other critical resources for agricultural production, are, in the Kalahari context, closely connected and therefore will be considered together. Kalahari land is not ideal farming land but it is usable once sufficient labour has been invested in it; it is non-commodified; and it is readily available.[19] Any adult should be able to obtain a plot. If, after approximately

---

[19] The fact that agricultural land is not commodified in Botswana with the exception of a few freehold blocks, is significant, and amongst the Kgalagadi has had a limiting effect on the degree of socio-economic inequality. It has contained, within certain limits, the distinctions between locally based patrons and clients; and it has prevented the growth of a landless agricultural labour class (the argument could be made that the Sarwa constitute such a class, but that is a complex issue beyond the scope of this chapter (cf. Wilmsen 1989)). It has also contributed to flexibility and diffuseness in patron–client bonds in that clients are less likely to be bound, through land, to any particular patron. Thus, although the commercialization of agro-pastoralism and its influence on social differentiation in the Kalahari bears strong resemblances to patterns that have existed elsewhere, there are important differences that mute certain common effects (cf. Scott 1972).

five years, the plot has been unused, the claim can be challenged. But the resources, largely labour, to clear, fence, and plough land are in short supply. Thus although 20-hectare plots are the standard land board allocation, very few farmers actually use the full extent of their fields.

It is necessary here to disaggregate the variable of labour along gender lines because what is in abundance is the labour associated with women—weeding, bird scaring, harvesting, crop preparation. It is the labour associated with men—clearing and fencing fields, the training and supervision of oxen—that is in short supply. Thus Batswana have evolved a number of strategies that allow extra household access to prepared fields. Many of these arrangements include access to draught-power and the associated labour, and most entail a labour but not cash exchange. In the 1970s most households with well-prepared fields allocated subplots regularly; in a year of late rains the number of people requesting subplots rose dramatically. For example, in 1979 the rains came late; one prominent farmer allocated nine plots within his fields to various people, mostly single mothers.

Similar to oxen-sharing arrangements these arrangements can be placed on a continuum of egalitarian to hierarchical with some amounting to simple field-sharing (not share-cropping) and others resembling—but falling short of—wage labour arrangements. Amongst the least egalitarian is a system referred to as *majako* or 'putting in hands'. In the case of *majako*, a subplot is rarely granted. Typically, *majako* involves helping in the fields, particularly with the weeding and harvesting (seldom with the ploughing and planting) and receiving as payment a specified portion of the harvest.

Systems of field-sharing have come under pressure; tractors make it easier for people to plough the full extent of their cleared fields. Increasing commercialization has influenced some farmers to sell a portion of their harvest. However, commercialization of the arable sector has not followed at the same speed as that of livestock; arable agriculture is, for a variety of reasons, less profitable. In addition, well-endowed farmers are less inclined to dispense with the primary benefit provided to them by field-sharing and that is labour. Agriculture, as it is practised in the Kalahari, is more labour intensive than pastoralism. Thus, while the greatest blow to oxen sharing arrangements has come from commercialization, with the state hastening the process, for field-sharing arrangements the greatest blow has come directly from the state. A brief description of the most radical of the agricultural subsidies will illustrate this point.

The ARAP (Accelerated Rain-fed Arable Programme) was announced in late 1985. It was formulated by high-level government fiat and was not widely discussed or evaluated prior to being instituted. In fact, most members of the Ministry of Agriculture first heard about the programme on the radio at the same time as everyone else. This was atypical for Botswana and demonstrated that the government was also utilizing the drought as a licence for policy-making methods that would have been unacceptable in a non-crisis situation.[20] ARAP was meant to be a one-off, one-year affair but instead lasted six years, and some farmers employed it repeatedly. Agricultural demonstrators came to the villages and held meetings to explain ARAP. It sounded too good to be true: farmers would be paid the equivalent of P50 per hectare for every one they ploughed up to ten hectares and they would receive additional funds for properly destumping and weeding their fields, for using fertilizer, a planter, and so on. Yet ARAP, like the other schemes, was under-utilized.[21] Why?

The agricultural subsidies required users to have fields registered in their own names, thereby excluding all people employing field-sharing methods. This attempt to rationalize agricultural organization, was perhaps well intended, but it put the subsidies beyond the grasp of a significant proportion of rural residents. Many of the poorer peasants, particularly from female-headed households (who comprise over one-third of rural households (Izzard 1979)), were, *de facto*, defined as ineligible. There is no Botswana law that prohibits women from having land but customary land-tenure arrangements are such that areas of agricultural land are associated with groups of agnatically related kin; the normative pattern is for women to gain access to land as wives. While the reality of a very high number of unwed mothers renders this ideal unworkable, custom is powerful and it is not always easy for women to apply for land. *And*, as I have emphasized, even if women have their own fields, as some do, their capacity to mobilize the resources necessary, largely male labour, to use the land is circumscribed. Thus the various arrangements formulated by rural peoples have permitted women to produce subsistence for their families while, at the same time, minimizing the costs and difficulties involved including the ones entailed in too overtly violating social norms.

It is ironic to note that ARAP subsidies actually appear to have promoted the least egalitarian sort of field-sharing arrangements, *majako*, while limiting field-sharing opportunities that grant greater autonomy to extra-household field users.

'Liberation' or New Patrons for Old

In 1986, at a talk I attended in Botswana's capital, a Ministry of Agriculture official, questioned about the fact that his ministry's schemes undermined informal rural-based land and draught-power sharing arrangements,

[20] For example, ALDEP (Arable Lands Development Programme), an agricultural subsidy programme not tied to the drought, was introduced in 1981 after years of planning and discussion and a two-year pilot phase.

[21] Eventually they were more widely used but still disproportionately by the wealthier farmers. ARAP's terms changed over the years and was very expensive for the government, accounting for 48% of total non-recurrent drought expenditures in 1988–9 (Valentine 1993: 117).

responded in a dismissive manner—'We want to liberate people from all those sorts of relations,' implying that they were nothing more than patron–client relations characteristic of a backward nation. What was apparent in his view and in both the discourse and policy of development planners was an attempt to project upon Botswana's rural populace an ideal of individualized nuclear family production units that functioned independently of one another but in conjunction with the state. Horizontal (in terms of inter-household, not necessarily in terms of equality) linkage should be minimized but vertical linkage to the state encouraged, initially in the form of aid but eventually through greater involvement in credit schemes, marketing, taxes, etc. To some extent such methods of attempting to restructure and exert control over the rural areas according to an imposed and idealized Western model simply follow in a long line of practices such as census-taking, tax collecting, land registration, and aid schemes, which have emerged from similar ideologies and have had similar effects.

In instituting the government subsidy schemes (and, indeed, all the drought relief measures) greater infrastructure was established—new agricultural demonstrators were hired, new offices created, communication and transport networks were enhanced, more farmers were registered, fields measured, and so on. Such increased presence of the state and state knowledge of rural people's economic situation and activities is surely a double-edged sword. It does enable the delivery of expanded services but, at the same time, it facilitates greater control and wielding of power through the state's growing bureaucracy; it is an instrumental effect of 'the development apparatus' (Ferguson 1990).

Government policy-makers want to 'liberate' rural households from relations that keep them beholden and dependent on others. While in the minds of the planners this might have been a noble sentiment it failed to acknowledge the realities of poverty and interdependence among the majority of rural dwellers. And it failed to recognize the good sense that many of these arrangements make in the context of a largely resource-poor rural populace attempting to minimize the risks inherent in producing in a marginal environment.

It is true that many of the rural-based entitlement relations are premised on inequality, and serve to perpetuate that inequality. I do not mean to celebrate or romanticize the situation, only to note that the relations do allow wide-scale production. And, as I have already noted, the majority of them follow a certain cultural logic whereby the poorer member's productive potential is enhanced, to some extent, by the relation. They are not simply the sort of wage-labour arrangements that the Batswana say simply use up a person and do nothing to contribute to their own self-development (Comaroff and Comaroff 1987).

What appears to be happening in the present situation is not the ending of dependence, rather it is a shift: dependence on the wealthy is being replaced with dependence upon the state, one sort of paternalism for another.

Dependence is being bureaucratized and 'modernized'. And the degree of dependence being engendered is deeper than it appears. The alternative, at the moment, in spite of Botswana's economic expansion, is *not* one where the majority of the population have the opportunity to gain satisfactory employment which could offset rural underproduction. As the massive subsidies are withdrawn, what will happen? How will people plough if there are too few oxen? Will they mortgage their harvests to tractor-owners? Will the rural areas be more vulnerable to future drought as a result of recent state interventions? Questions such as these are endless and all revolve around the fact that the set of entitlements that allowed people to survive previous droughts and that allowed the majority of people to produce in good years have been undermined to the point that they no longer provide basic social security.

Locally based entitlements have been undermined on two fronts: the state, with its image of 'rationally based individualized peasant production', formulated a development policy that discouraged inter-household production relations. On the other and more fundamental front, the vitality of reciprocal use entitlements has been undermined as a result of long-term structural changes tied to commodification, privatization, and class formation.

Thus far the state has done well by its relief efforts; it prevented starvation, received international acclaim, and substantiated its own power-base. Indeed, drought relief has been established as the single most important reason people give for supporting the current government (Molutsi 1989: 128). The state has fulfilled some of its patronage duties and its clients have reciprocated with votes. But in averting famine has the state simply delayed the full effects of the 'crisis of social reproduction'? Is the state's position sustainable?

Conclusion

It is not drought that produces a crisis of social reproduction, the drought only hastens and renders visible what had been up to a certain point largely latent processes. Yet, at the same time, because of the severity of the externally wrought conditions, the crisis can be blamed on the drought, thus obviating the necessity to examine or acknowledge underlying causes. In this manner drought functions as a means of concealing such systemic realities, both for local farmers and for government bureaucrats. In addition, it is the perfect opportunity for breaking taboos, norms, and moral standards; wealthy cattle-owners could deny kin-based claims to access to their property with licence, and the state could institute policy in an unprecedented manner.

The observation of social activity around a drought is also revelatory for social scientists, both for what it reveals of social process and for the questions it poses for comparative purposes. The Botswana case is instructive. A relatively wealthy benevolent state set about to avert famine, a task at which it succeeded when others around it failed. However, in shifting entitlement to itself the state contributed to a parallel process of disentitlement. Is there necessarily a contradiction between state-based and local-based entitlements? How can they be compatible?

Similarly, such questions lead one to wonder about the relative advantages and disadvantages of different forms of patronage relations in varying contexts. What is entailed for a poor person in being a client of the state rather than of a rich neighbour in terms of getting access to productive resources and other needs of daily life? Wolf (1966: 86), following Pitt-Rivers (1954: 140), speaks of patron–client relations as 'lop-sided friendships' that involve some degree of trust and affect, and Scott (1985) emphasizes the importance to the poor of the élite being within their same moral community and thus within moral reach. Is the state within moral reach? Do its agents have the inclination or capacity to respond to the myriad needs that may emerge in a relationship as multifaceted as locally based patronage relations tend to be? Has sufficient state infrastructure—medical, educational, social security, transport, etc.—been established to replace the need for diffuse patron–client relations for the rural poor? Where and when is one form of patronage preferable to the other and what choices do peasants have in choosing patrons? What is involved when types and sources of entitlements transform and how is the liberatory quality of any to be assessed? My analysis alone does not answer such questions but points to the necessity of considering them in any analysis of rural change and development.

Postscript—1999

In Botswana 1999 was an election year; it was also officially declared a drought year. The relation between these two events is not causal but nor is it entirely casual. Since 1980, with one exception, every year has been officially declared a drought year, at least in certain parts of the country. Rainfall has varied widely and several 'good' years have been reported, but arable production for all crops continues to fall at a rate of approximately 3 per cent per annum (Republic of Botswana 1997: 237).[22]

The drought-as-culprit perspective is under attack; policy-makers are being implored to examine the structural determinants of poverty and to plan accordingly (Jefferis 1997).[23] Moreover, it is increasingly recognized that poverty alleviation and the delivery of many basic social and economic services take place largely under the guise of drought relief (*Mmegi*, 6–12 August 1999). Yet, for the time being, it appears that the state is still unable to dispense with drought.

---

[22] This decline is especially striking in the light of the heavy subsidization of this sector and the growing use (with government support) of donkeys as draught-power by many poor rural households.

[23] Jefferis is a member of the Botswana Institute of Development Policy Analysis (BIDPA) and his views reflect their important studies on poverty.

Acknowledgements

I wish to thank Angelique Haugerud, Ann McDougall, Michael Lambek, John Comaroff, Jim Scott, and Deborah Kaspin for their comments and the participants at the Trent University Comparative Development Studies Seminar and the Yale University Program in Agrarian Studies where I presented papers which were earlier versions of this chapter. I also wish to thank the Government of Botswana, and the people of Western Kweneng for allowing me to conduct research there, the Botswana Ministry of Agriculture—especially Yvonne Merafe—for facilitating my research, and the Ontario Graduate Scholarship and University of Toronto Open Fellowship for supporting my early fieldwork. The above bear no responsibility for the conclusions reached in this chapter.

References

ALVERSON, H. (1978), *Mind in the Heart of Darkness* (New Haven: Yale University Press).

BERRY, S. (1984), 'The food crisis and agrarian change in Africa', *African Studies Review*, 27/2: 59–112.

BONTE, P. (1975), 'Cattle for God: An attempt at a Marxist analysis of the religion of East African herdsmen', *Social Compass*, 22/3–4: 381–96.

——(1981), 'Ecological and economic factors in the determination of pastoral specialization', in J. Galaty and P. Saltzman (eds.), *Change and Development in Nomadic and Pastoral Societies* (Leiden: Brill), 33–59.

BUSH, R. (1988), 'Hunger in Sudan: The case of Darfur', *African Affairs*, 87/346: 5–24.

CARSTENS, P. (1983), 'The inheritance of private property among the Nama of Southern Africa reconsidered', *Africa*, 53/2: 58–70.

COLCLOUGH, C., and MCCARTHY, S. (1980), *The Political Economy of Botswana* (Oxford: Oxford University Press).

COMAROFF, J. L. (1982), 'Class and culture in a peasant economy: The transformation of land tenure in Barolong', in R. Werbner (ed.), *Reform in the Making* (London: Rex Collings, 85–113).

COMAROFF, J. L., and COMAROFF, J. (1987), 'The madman and the migrant: Work and labor in the historical consciousness of a South African people', *American Ethnologist*, 14/2: 191–209.

DREZE, J. (1990), 'Famine prevention in Africa: Some experiences and lessons', in J. Dreze and A. Sen (eds.), *The Political Economy of Hunger* (Oxford: Clarendon), ii. 123–72.

DREZE, J., and SEN, A. (1990), *The Political Economy of Hunger* (Oxford: Clarendon), ii.

FERGUSON, J. (1990), *The Anti-Politics Machine: 'Development', Depoliticization and Bureaucratic Power in Lesotho* (Cambridge: Cambridge University Press).

FRANKE, R., and CHASIN, B. (1980), *Seeds of Famine* (Montclair, NJ: Allanheld, Osmun).

GALATY, J. (1981), 'Introduction: Nomadic pastoralists and social change—processes and perspectives', in J. Galaty and P. Saltzman (eds.), *Change and Development in Nomadic and Pastoral Societies* (Leiden: Brill), 4–26.

GLANTZ, M. (1987), *Drought and Hunger in Africa* (Cambridge: Cambridge University Press).

GOOD, K. (1992), 'Interpreting the exceptionality of Botswana', *Journal of Modern African Studies*, 30/1: 69–95.

GRAY, R., and GULLIVER, P. (1964), *The Family Estate in Africa* (London: Routledge).

HAY, R. (1988), 'Famine incomes and employment: Has Botswana anything to teach Africa?', *World Development*, 16/9: 1113–25.

HITCHCOCK, R. (1987), 'Socio-economic change among the Basarwa in Botswana: An ethnohistorical analysis', *Ethnohistory*, 34/3: 219–55.

HOBSBAWM, E. J., and RANGER, T. (1983), *The Invention of Tradition* (Cambridge: Cambridge University Press).

HOLM, J., and MORGAN, R. (1985), 'Coping with drought in Botswana: An African success', *The Journal of Modern African Studies*, 23/3: 463–82.

IZZARD, W. (1979), *Rural–Urban Migration of Women in Botswana* (Gaborone: Central Statistics Office).

JEFFERIS, K. (1997), 'Poverty in Botswana', in D. Nteta, J. Hermans, and P. Jeskova (eds.), *Poverty and Plenty* (Gaborone: Botswana Society), 33–58.

KUPER, A. (1966), 'Kinship and politics in a Kgalagari village', Ph.D. Thesis, Cambridge University.

——(1970*a*), *Kalahari Village Politics* (Cambridge: Cambridge University Press).

——(1970*b*), 'The Kgalagari and the jural consequences of marriage', *Man*, 5/3: 466–82.

LOFCHIE, M., and CUMMINS, S. (1982), 'Food deficits and agricultural policies in tropical Africa', *The Journal of Modern African Studies*, 20/1: 1–25.

Mmegi (newspaper), 6–12 August 1999, 'Researcher spells out issues for the elections'.

MOLUTSI, P. (1989), 'Whose interests do Botswana's politicians represent?', in J. Holm and P. Molutsi (eds.), *Democracy in Botswana* (Gaborone: Macmillan), 120–32.

ORTNER, S. (1989), *High Religion* (Princeton: Princeton University Press).

PETERS, P. (1984), 'Struggles over water, struggles over meaning: Cattle, water and the State in Botswana', *Africa*, 54/3: 29–49.

——(1992), 'Manoeuvres and debates in the interpretation of land rights in Botswana', *Africa*, 6/3: 413–34.

PITT-RIVERS, J. (1954), *The People of the Sierra* (London: Weidenfeld & Nicholson).

REPUBLIC OF BOTSWANA (1989), *Budget Speech* (Gaborone: Government Printer).

——(1991), *National Development Plan* (Gaborone: Government Printer).

——(1993), *Budget Speech* (Gaborone: Government Printer).

——(1997), *National Development Plan* (Gaborone: Government Printer).

SAHLINS, M. (1972), *Stone Age Economics* (Chicago: Aldine).

SCOTT, J. (1972), 'The erosion of patron–client bonds and social change in rural Southeast Asia', *Journal of Asian Studies*, 32/1: 5–37.

——(1985), *Weapons of the Weak* (New Haven: Yale University Press).

SEN, A. (1982), *Poverty and Famines: An Essay on Entitlement and Deprivation* (Oxford: Clarendon).

SHIPTON, P. (1990), 'African famines and food security: Anthropological perspectives', *Annual Review of Anthropoloqy*, 19: 353–94.

SOLWAY, J. (1979), *People, Cattle, and Drought*, Rural Sociology Unit Report (Gaborone: Government Printer).

——(1987), 'Commercialization and social differentiation in a Kalahari village, Botswana' Ph.D. dissertation, University of Toronto.

SOLWAY, J., and LEE, R. B. (1990), 'Foragers, genuine and spurious: Situating the Kalahari San in history, *Current Anthropology*, 31/2: 109–46.

SWIFT, J. (1993), 'Understanding and preventing famine and famine mortality', *IDS Bulletin*, 24/4: 1–16.

VALENTINE, T. (1993), 'Drought, transfer entitlements and income distribution: The Botswana experience', *World Development*, 21/1: 109–26.

VAUGHAN, M. (1987), *The Story of an African Famine* (Cambridge: Cambridge University Press).

WATTS, M. (1987), 'Drought, environment and food security: Some reflections on peasants, pastoralists and commoditization in dryland West Africa', in M. Glantz (ed.), *Drought and Hunger in Africa* (Cambridge: Cambridge University Press), 171–211.

WILMSEN, E. (1989), *Land Filled with Flies: A Political Economy of the Kalahari* (Chicago: University of Chicago Press).

WOLF, E. (1966), *Peasants* (Englewood Cliffs, NJ: Prentice Hall).

WORLD BANK (1992), *World Development Report 1992: Development and the Environment* (New York: Oxford University Press).

WYLIE, D. (1989), 'The changing face of hunger in Southern Africa', *Past and Present*, 122: 159–99.

10

Local Lessons for Global Problems: Contributions to Global Debates

Deborah Sporton and David S. G. Thomas

Introduction

The Kalahari has often been romanticized in the popular Western imagination as one of the last wilderness environments inhabited by 'first peoples' whose livelihoods, sustained by subsistence practices, are symbiotically attuned to the harsh natural environment (see for example, van der Post 1958). This interdisciplinary collection has drawn together a range of examples that highlight the Kalahari today as a landscape criss-crossed with fences, populated by cattle, and inhabited by 'real' people for whom contemporary enviro-social relations and livelihood outcomes are both embedded within a long and complex history of social and environmental change and increasingly subject to the commercial pressures of the global economy.

While this collection provides, on the one hand, a comprehensive overview of the dynamics of livelihoods and environment in the Kalahari, it has also highlighted key enviro-development issues that resonate across the individual chapters and that have relevance to wider global debates. In this concluding chapter, themes have been identified that cross-cut individual chapters, that are paralleled in other regions, and that have wider relevance to global concerns that seek to ensure sustainable livelihoods in climatically marginal environments.

Political Economy

The case of the Kalahari has highlighted the significance of political economy to understanding the dynamics of livelihoods and dialectical enviro-social relations. The Kalahari environment has for centuries been subject to competing and overlapping claims over land and natural resources. As Deborah Sporton and Chasca Twyman reveal in Ch. 3 this stems from pre-colonial ethnic tensions and hegemonies endorsed through colonial rule and in the post-colonial politicization of ethnicity in South Africa and Namibia through apartheid. As a result both control over and access to natural resources have been privileged to certain groups and denied to others, creating a growing underclass of rural poor constituted largely from 'first peoples'—for example, Basarwa groups in Botswana and the Damara of Namibia. Such historically rooted, racialized systems of social relations underpin access to livelihood assets such that, without redress, policy interventions may provide only short-term solutions, concealing longer-term problems and exacerbating social polarization.

Within the Kalahari of South Africa and Namibia, for example, the inequitable distribution of land, the legacy of apartheid, favouring white commercial farmers has placed pressure on marginal communal lands. Policy attempts aimed at redistribution in the Kalahari of Namibia have bowed to commercial pressures to preserve the status quo with resettlement taking place within the communal sector. The current situation in Zimbabwe provides a timely example of how such commercially driven policies can backfire in the longer term.

A long history of conflict and territorial appropriation within the Kalahari has created a situation whereby the majority of rural dwellers lack title over resources and land and the means to sustain livelihoods. In their study of the TGLP, Deborah Sporton and David Thomas (Ch. 6), for example, highlight the ways in which the policy failed to recognize the diverse and competing claims to land and natural resources among rural populations. Historically, the movement of hunter-gatherer and pastoralist populations within territories constituted a traditional response to seasonal and drought-induced shortages of water and food. With the advent of cattleposts important clientage relationships had developed largely between Basarwa populations and the 'owners' of water resources as grazing reduced the availability of veld foods. Underlying this relationship was the belief among local populations that the Kalahari was *their* land and that cattlepost-owners were merely using *their* land. The ranch system undermined this relationship by dispossessing traditional owners thereby reducing many residents to the status of squatters whose grazing and cultivation rights had been removed.

Both Chasca Twyman (Ch. 7) and Robert Hitchcock (Ch. 8) highlight that even in Wildlife Management Areas (WMAs) and conservancies, established at the time of TGLP to provide alternative livelihood opportunities for non-livestock-owners, ethnicity constitutes a key dimension affecting natural resource entitlements. Chasca Twyman's study of the Ghanzi WMAs highlights the ways in which Basarwa populations, who constitute the majority within these settlements, have no title to the land or resources and therefore lack power to exclude powerful Batswana and Bakgalagadi cattle-owners encroaching on their resources and distorting their entitlements. While Wildlife Management Areas have been applauded for their environmental goals they have contributed to the further marginalization of many rural dwellers.

Jacqueline Solway (Ch. 9) discusses the way in which drought has been discursively constructed as a crisis to mask social and economic problems—the solution by government to deal with this short-term crisis through drought relief measures in fact exacerbated social and economic inequality. By describing drought as a revelatory crisis, she identifies how economic relations of production were transformed, contributing to further polarization along ethnic lines. Although ethnicity, as a social category, is no longer recognized by the state and downplayed in official discourse, with all citizens recognized as equal Batswana, the hierarchical system of ethnic social relations established in pre-colonial times remains an important, albeit shifting, social construct that structures access to livelihood assets (see Motzafe-Halle 1998).

The case of the Kalahari reveals how political economy and entrenched ideologies have structured resource relations, capabilities, and livelihoods within the region. The challenge for development policy and successful environmental mangement is to recognize the historically rooted power-relations underlying contemporary enviro-social relations.

Vulnerability and Risk

Vulnerability and risk are key elements affecting the sustainability of livelihoods (Scoones 1998) and are vital to understand if poverty reduction is to be effected. Vulnerability exposes people, often the poorest members of society, to the negative dimensions of risk usually because of their lower and less diversified capital asset status that they can fall back on at times of risk: they live closer to the edge. People may, however, develop elaborate mechanisms for coping with risk, for example through mobility, and support from social networks (Bryceson 2000). These mechanisms, however, do not necessarily eliminate vulnerability, and may offer short-term protection at the cost of longer-term benefits (e.g. Bebbington 1997), such that poverty is ultimately increased.

Within the Kalahari, drought and rainfall variability represent a risk to the livelihoods of the most vulnerable and as such have constituted a pervasive theme throughout the chapters of this book. From the perspective of livelihoods, it is not climatic variability *per se* that is the primary problem, but the risks that the variability induces, and people's vulnerability to the effects of these risks, that are more critical (Blaikie *et al.* 1994, Ribot *et al.* 1996). Although forecasted increases in rainfall variability associated with global warming are unlikely to have a marked effect on the Kalahari (see Ch. 2), the experience of the Kalahari provides lessons that have salience for other dryland regions both now and into the future. In particular, what are the risks (both direct and indirect) to livelihoods presented by drought; how, if at all, do people mitigate the effects of drought, and who are the most vulnerable to these risks?

Drought and environmental degradation are terms that have until recently been widely used in a related way in explanations of the relationships between dryland people and their environments. In Ch. 2, David Thomas highlights current thinking that challenges not only the perception of major ecological changes as degradation but also their irreversibility. Rather, emphasis is placed upon the resilience of the Kalahari ecosystem to recover from drought. Policy interventions promoting further sedentarization, and fragmentation of the land, rather than drought *per se*, pose the key threats to the long-term viability of the natural resource base of the Kalahari environment into the twenty-first century. This fragmentation may reduce severely the opportunities for environmental recovery from disturbances, whether these disturbances are natural or human-induced. However, the nature of environmental changes and their spatial and temporal scales are complex in the Kalahari. Andrew Dougill (Ch. 5) reveals how drought along with fire actually restricted bush encroachment in the eastern Kalahari and may in fact be a resilience rather than a degradation mechanism. Given the nutrient-poor status of Kalahari soils under natural conditions, ecological changes in the Kalahari may not impact on the soil in the manner that may occur in areas of richer soil resources. The studies that Dougill reports found no evidence of a link between soil and vegetation changes, suggesting that natural events such as drought or fires, or management actions such as prescribed burning, may enable a return of bush-dominant areas associated with degradation to a grass-dominant state. This suggestion needs further investigation with regard to long-term ecological monitoring programmes following major drought and fire events. The implications with regard to the future sustainable use of Kalahari rangelands are far-reaching and should be incorporated into future land-use strategies.

Kalahari populations have been found to employ a range of strategies and flexible adaptations to cope with risk associated with drought and rainfall variability. Robert Hitchcock (Ch. 9) highlights a number of strategies that have traditionally been employed by northern Kalahari communities to mitigate the effects of drought and livestock disease. These have included labour intensification to maximize returns from scarce resources, loaning livestock to others away from drought areas through the *mafisa* system, increased storage of and scavenging for meat, crops, and veld foods. Jural restrictions on resource use to prevent depletion were also imposed during shortages. By relying on indigenous knowledge, drought-resistant crops were planted and soil fertilization and water conservation strategies employed. The role of population mobility between group territories linked through networks of social relations has historically allowed both groups and individual herders with cattle to move to areas where natural resources and grazing were more abundant. With increasing privatization and new fencing policies this strategy for example in the eastern Kalahari (see Ch. 6) is no longer a possibility. Indeed Deborah Sporton and David Thomas discuss how the allocation and demarcation of leasehold ranches under the TGLP has disrupted the social and spatial mobility networks of Basarwa populations in particular. Ironically, while curbing potential drought coping strategies amongst the most vulnerable groups, the privatization of the range through policies such as the TGLP and the New Agricultural Policy has facilitated new grazing strategies among élite leaseholders. During times of drought ranch lessees were found to exercise their dual grazing rights moving cattle onto already pressured communal areas. Chasca Twyman (Ch. 7) also identifies similar strategies within the Ghanzi WMAs where local farmers, having depleted their own grazing resources, are moving into WMAs pressuring the natural resource base of the local communities.

Jacqueline Solway (Ch. 9) argues that sharing relations between kin have traditionally helped to offset risk and vulnerability. However, the progressive privatization and consolidation of economic resources in response to policy has led to the creation of a rural élite whose withdrawal of such resources from the communal domain has deprived poorer sections of the community of their use. In the past the effects of drought were mitigated by ploughing later if rains were late, and among poorer households by cutting their losses and farming plots in fields owned by those who had greater access to draught-power. Access to the livestock of larger farmers for milk and draught-power, through the *mafisa* system, sustained the livelihoods of smaller farmers. With the increasing privatization of the range and commercialization of agricultural production, the large farmers who constitute the rural élite are less willing to share access to these economic resources, resulting in increasing social polarization and vulnerability to the effects of drought. This situation, she argues, has been encouraged by the government in Botswana, which through subsidies and drought relief programmes has further eroded horizontal (inter-household) linkages in favour of vertical linkages, and dependence on the state. This, she contends, fails to acknowledge the significance of interdependence among rural households in the mitigation of vulnerability to drought. Sporton and Thomas (Ch. 6) highlight the ways in which rural populations nevertheless still draw upon these horizontal linkages in times of livelihood stress. For example on certain TGLP ranches in the Ncojane area of western Botswana, fences were removed to allow the free movement of livestock and the sharing of water in line with the traditional communal-based cattlepost system. On other fenced ranches, the livelihood assets of workers and residents (wages, access to milk and meat from cattle, veld products, and wildlife) was improved and vulnerability reduced under management systems that emphasized reciprocity.

The significance of drought relief and state intervention has impacted all spheres of livelihoods. Heralded as an African success story, Botswana's drought relief programme has undoubtedly proved a successful short-term safety-net preventing the worst ravages of drought. However, its utility as a model and a lesson for other drought-prone areas is questionable. Considerable evidence from the chapters in this volume suggest that social security entitlements are threatening the longer-term sustainability of livelihoods. While state intervention has undermined traditional patronage relations and social capital it has also created a dependency culture. The livelihoods of a growing proportion of the rural population have become dependent on cash, food, and labour-based relief such that many no longer have the means to sustain livelihoods during years of reasonable rains. As a result the annual declaration of drought (and hence the opening of the drought relief programme) in Botswana is increasingly regarded as having little to do with the rains but as a political act aimed at appeasing the rural populace for whom 'drought' ironically now mitigates vulnerability to hardship and poverty.

Discourses of Environmental Management

Underlying many policy initiatives in the Kalahari has been the concern to reduce perceived environmental degradation caused by the expansion of livestock into a drought-prone environment. These interventions, for example the TGLP, have been informed by dominant Western scientific discourses (discussed by David Thomas and Andrew Dougill in Chs. 2 and 5) that have until recently associated bush encroachment around boreholes with a permanent reduction in the available grass resource, and as a manifestation of desertification. These scientific 'wisdoms' have now been replaced by those that recognize the resilience of dryland ecosystems and suggest that with time and effective management, rangeland ecological changes can be reversed, within the natural complexity of state and transition behaviour that is dominant in dryland ecosystems.

Andrew Dougill highlights the potential of fires (natural or managed) to return the range ecosystems to a grass-dominant form. However this is not a new strategy but is one that Jeremy Perkins *et al.* (Ch. 4) point out has been used for centuries by hunter-gatherers in the Kalahari (and aboriginal populations in Australia) to maintain optimum rangeland habitats for wildlife. Perkins argues, however, that this traditional management strategy has been ignored in the literature as it has hitherto lacked scientific backing. Indeed, managed fires were banned in the Kalahari during the 1980s.

Although the TGLP was commercially driven there was an underlying environmental rationale informed by this 'desertification' discourse, namely to reduce levels of degradation by introducing management strategies such as rotated grazing, fencing ranches to ensure cattle could be effectively and exclusively managed, and regulating cattle numbers. Deborah Sporton and David Thomas discuss the complex and spatially variable interactions between ranch management and the natural variability characteristic of the semi-arid ecosystem. Although there was strong evidence that bush encroachment was occurring under TGLP there was marked variability in its extent between and within ranch blocks. In the Ncojane ranch block of the western Kalahari, severe drought precluded the operation of range management strategies that were inflexible, such that lessees reverted to traditional cattle-post systems of livestock management, thereby safeguarding the livelihoods of ranch residents. In the eastern Kalahari, TGLP ranches were not drought-affected but similar levels of degradation resulted through lack of management to the detriment of local livelihoods.

Science remains the dominant discourse shaping our understanding of the environment. While the focus of debates within environmental science is now to question the permanence, or irreversibility, of degradation, the significance to livelihoods of local practices that challenge even current environmental orthodoxies is highlighted in several chapters. In short, people are making a living and coping in times of crisis by doing things that we least expect, that are 'scientifically' irrational.

Global Markets vs. Local Needs: Policy Challenges for the Twenty-First Century

Although global agendas now encourage grass-roots participatory approaches to poverty alleviation and the creation of sustainable livelihoods (CCD, World Bank 2000), the reality of policy implementation in the Kalahari region reflects conflicting pressures to commercialize national economies.

Within Botswana the actions of multilateral donor agencies have fundamentally influenced national government policies that bear upon both livelihoods and the Kalahari environment which sustains them. The expansion of commercial pastoralism in the Kalahari, for example, has been driven since 1975 by the European Union's Lomé Convention, in particular the *Beef and Veal Protocol 6*, which abolished import tariffs and has provided an export market for beef at preferential prices. Botswana now exports 80 per cent of beef production to the EU, which has fixed prices 50 per cent higher than world prices. As several of the chapters attest, this external pressure has contributed to a suite of top-down government policies and directives that have emphasized national rather than local priorities, the concerns of the national economy over local livelihoods and the environment.

Several chapters (see e.g. Jeremy Perkins *et al.* Ch. 5) highlight the deleterious impact on wildlife populations of veterinary cordon fences erected in response to EU disease control regulations—a situation mirrored in other wildlife-rich countries such as Zimbabwe. Periodic outbreaks of Foot-and-Mouth Disease, and more recently Contagious Bovine Pleuropneumonia, have threatened livestock exports and have thus been dealt with swiftly by government (See Ch. 8) to the detriment of local livelihoods.

The Tribal Grazing Land Policy, driven by Lomé, destabilized enviro-social livelihood systems to the extent that further policies were introduced to accommodate those displaced by the policy—WMAs and Remote Area Development (RAD) Centres were established. As Chs. 3, 6, and 7 reveal, these new livelihood options have proved problematic and, in the case of RAD service centres, unsustainable. Throughout, the consultation of local communities in policy implementation has been negligible, key issues of entitlement to land and resources have been ignored. Indeed, commodification and commercialization of cattle production have transformed the social relations around which livelihoods were organized, creating a growing underclass of rural poor.

The replacement of Lomé by a Generalized System of Preferences (GSP) is being introduced gradually over five years, during which time preferences may be maintained through economic partnership agreements. However, a proposed 30 per cent cut in the intervention price of beef will lead to a corresponding loss in export earnings with knock-on effects for policy at the national level and the viability of certain commercial ranches at the local level. It remains to be seen how the removal of this market distortion will affect the peoples and environment of both the Kalahari and of other regions affected by Lomé. On the positive side, under the new GSP system a differentiated, incentivized approach is to be introduced in favour of countries applying social and environmental policies in line with international agreements. For example, in Botswana over 12 m. ECU (approximately one-third of the total EU development fund to Botswana) has been allocated to natural resource utilization and conservation with emphasis placed on support for policies and projects that encourage three actions: community-based wildlife management in the Kalahari, tackling the issue of wildlife depletion, and tackling the effects of cattle on rangeland degradation (EU and Government of Botswana 1996). However, this about-turn in policy means that the EU will now exert direct, rather than indirect, control over national government policies, which will need to be more responsive to EU priorities and requirements than before.

Concluding Remarks

This analysis demonstrates that the Kalahari is a complex politico-socio-environmental matrix, where the demands and aspirations of international, national, and local policies and preferences have exerted, in an ever-increasing way since the mid-twentieth century, complex and often conflicting influences on the people of the region. Many actions at the international and national levels affecting Kalahari countries, particularly Botswana which has been the focus of much of the research reported in this book, have created situations that have disadvantaged traditional sustainable livelihood activities. As the pressures to Westernize and internationalize have grown upon the countries of the developing world, it would appear that forms of neo-colonialism and internal social and economic polarization have grown. Many actions impacting on the Kalahari, driven by external and national priorities, have effected outcomes that appear unsustainable in terms of elements of the environment and the livelihoods of rural people. Strategies that developed over centuries, even millennia, to cope with the environmental uncertainties of this dryland area have been disabled in a matter of a few decades. Attempts to re-establish elements of these strategies, in the hope of improving rural livelihoods and the environment, will have to be grounded in more than rhetoric and goodwill if they are to succeed, especially with the additional layer of growing environmental and climatic uncertainty in the twenty-first century.

References

BEBBINGTON, A. (1997), 'Social capital and rural intensification: Local organizations and islands of sustainability in the rural Andes', *Geographical Journal*, 163: 189–97.

BLAIKIE, P., CANNON, T., DAVIS, I., and WISNER, B. (1994), *At Risk: Natural Hazards, People's Vulnerability and Disasters* (London: Routledge).

BRYCESON, D. (2000), 'Rural Africa at the crossroads: Livelihood practises and policies', *ODI Natural Resource Perspectives*, 52.

MOTZAFI-HALLER, P. (1998), 'Beyond textual analysis: Practice, interacting discourses, and the experience of distinction in Botswana', *Cultural Anthropology*, 13/4: 522–47.

RIBOT, J. C., NAJAM, A., and WATSON, G. (1996), 'Climate variation, vulnerability and sustainable development in the semi-arid tropics', in J. C. Ribot, A. R. Magalhaes, and S. S. Panagides (eds.), *Climatic Variability, Climatic Change and Social Vulnerability in the Semi-arid Tropics* (Cambridge: Cambridge University Press), 13–51.

SCOONES, I. (1998), 'Sustainable rural livelihoods. A framework for analysis', *IDS Working Paper*, 72 (Brighton: IDS).

VAN DER POST, L. (1958), *The Lost World of the Kalahari* (London: Hogarth).

WORLD BANK (ed.) (2000), *World Development Report 2000/2001: Attacking Poverty* (Oxford: Oxford University Press).

INDEX